The Still-Burning Bush

Stephen Pyne is an emeritus professor at Arizona State University. Among his many books are *Burning Bush: a fire history of Australia*, and *Fire: a brief history*.

For Sonja
there in spirit

THE STILL-BURNING BUSH

Stephen Pyne

SCRIBE
Melbourne • London

Scribe Publications
18–20 Edward St, Brunswick, Victoria 3056, Australia
2 John St, Clerkenwell, London, WC1N 2ES, United Kingdom
3754 Pleasant Ave, Suite 100, Minneapolis, Minnesota 55409, USA

First published by Scribe 2006
This updated edition published 2020

Typeset in 11/16pt Minion by the publishers
Printed and bound in the UK by CPI Group (UK) Ltd, Croydon CR0 4YY

Scribe Publications is committed to the sustainable use of natural resources and the use of paper products made responsibly from those resources.

9781950354481 (US edition)
9781922310309 (Australian edition)
9781925938494 (ebook)

A catalogue record for this book is available from the National Library of Australia.

scribepublications.com
scribepublications.com.au

Contents

Preface

When *The Still-Burning Bush* was published, 15 years had passed since I had written *Burning Bush: a fire history of Australia*. Now, after another 15 years, that extended essay is itself ready for reconsideration.

My original text focused on policy and practice and what had occurred since the 1983 Ash Wednesday fires that had concluded *Burning Bush*. Against the usual narrative of colonial disruption, I wanted to isolate what I regarded as a remarkable expression of continuity. The firestick became an organising device, a kind of pyric wizard's wand, for interpreting the human history of fire in Australia.

In the days of European exploration, parties could track the presence and movement of people across the landscape by their fires. The firestick left a record in smoke, ash, and green pick. So, too, one might track the movement of parties and ideas across the contemporary political landscape by firestick controversies with their legacy of words, symbols, and flame on the land. That, at least, was, and remains, the large ambition of this text.

Its theme pivots on the Australian firestick. Humans had tamed lightning into a firestick, and then successive settlers had reimagined and repurposed it to fashion landscapes that better suited their conception of who they were and where they were.

The firestick farming of Indigenous Australians morphed into a European firestick agronomy, then into a firestick forestry, and a firestick ecology. Throughout, the firestick remained a point of contact between people and country — in fact, the means and emblem of humanity's unique relationship to Earth's biota. Still, it makes an odd implement, less a physical tool than a catalyst for ecological process, and not so much an implement as a relationship that soft-welds people to place.

Each avatar sparked controversy, since each served a different worldview. After the Alpine fires, a vigorous debate sharpened between those who proposed to burn for hazard reduction and those who wanted to burn to advance ecological values. Firestick forestry argued that routine burning could reduce fuels and make fire protection easier; a fire-adapted nature would sort out the ecology. Firestick ecology argued that suitable burning could advance biodiversity and maintain critical processes, out of which nature would find a suitable fuel array. The differences may seem subtle, but they are real, and they have political consequences in deciding who should hold the firestick. The firestick served as a political lightning rod. *The Still-Burning Bush* tracked the origins and character of that controversy.

Typically, and not just in Australia, competing groups stand around a common fire but with their backs to the flames, each speaking to a separate group, using the fire to animate another agenda. Arguments over which firestick should rule take on the character of gang signs. Politics then picks up the firestick not as a device to discharge duty-of-care obligations to the environment, but as a club with which to beat down opponents.

Since the original edition, two spectacular outbreaks of bushfire have shifted the baseline for Australian fire history, the firestick has found a new iteration in cultural burning, and the deep driver of fire on Earth, the burning of fossil fuels, has seized the commanding heights of fire's narrative. Taken together, they would argue for a new synthesis organised around industrial combustion and how it has both broken and bolstered the inherited story.

The Black Saturday bushfires of 2009 and the Forever fires a decade later — a Red Summer of bushfires — have hammered two geodetic markers by which to triangulate into the likely future of Australian fire. They were not merely tragedies — although they were that, on a colossal scale — but national traumas. It was as though Australia had been visited by terrorist attacks, with the bush itself as the source of terror. They seemed to question the very premises of modern Australia, how a first-world economy and way of life might align with a land capable of such fury.

The political effects of even the most savage fires pass, often quickly, perhaps no longer than the digital half-life of a meme. But these fires have had an outsized cultural impact in lives lost, houses destroyed, the seeming futility of control efforts, the pervasiveness of burning in every part of the continent, and the immense, lurid smoke palls that smothered the city-states that house most Australians, especially Sydney, Melbourne, and Canberra. The fires burned on and on and on. The smoke extended the fires' reach far beyond the flames' grasp. It affected air quality in New Zealand. It coloured sunsets in South America. Eventually, it ringed the globe. Media attention followed, putting Australia's inextinguishable bushfires into front pages and twitter storms.

The fires also fused with global alarm over a fast-morphing climate powered by the burning of fossil fuels. Past inquiries into bushfires had focused on living landscapes. What will follow the Forever fires will also probe the burning of lithic landscapes — those reservoirs of fossil biomass that humans were exhuming and combusting. The sources of fuel have proved vaster than the sinks for their combusted by-products.

Something similar has happened with explanations: the range of impacts from fossil fuels is far broader than global warming suggests. Like a driverless car, bushfires integrate everything around them as they blast down the road. Climate change is acting as a performance enhancer, widening the seasonal opportunities for fires, quickening the tempo of their appearances, and intensifying the flames that occur. It affects not only wetting and drying and winds, but the biotic character of country — which is to say, the fuels available for bushfire. Much of the existing scene is the outcome of land use, fire practices, and the energy that runs Australia's economy. Here lies the second influence of fossil fuels: they underwrite much of how modern Australians live on the land. Even close-crowding fire control is only possible through pumps, chain saws, engines, aircraft, and lorries on roads themselves cut, graded, and perhaps paved through a reliance on fossil biomass. The burning of lithic landscapes is the magma chamber that underlies the eruptions that have plagued living landscapes in recent decades.

The shift to fossil fuels has sparked fire crises in developed nations everywhere. Those that were predisposed to fire have burned more alarmingly under the new dispensation. In the United States, for example, a fire crisis boiled over in the 1960s and 1970s from the misguided efforts to exclude fire. The problem was equally about fuels and ecology. Without reducing hazards, modest fires could become monsters; and without the catalytic jolt of burning, many biomes decayed. In 1968 the National Park Service and in 1978 the U.S. Forest Service committed to policies of fire reintroduction. The programs stalled as politics shifted in the 1980s, and was soon too little too late.

Forty years later, local fire crises had metastasised into a global fire epoch, a Pyrocene, in which the Earth was assuming the fire-informed equivalent of an ice age. Australia found itself on the front lines. Australia's fires, and Australia's fire discourse, would not stay in Australia any more than their smoke.

How living landscapes interact with lithic ones is not merely the great fire question of our age, but a primary shaper of our time. Its history has been full of paradoxes. We don't have more landscape fire than ever before; we have less. We see the fire orgies; we don't see the fire famines. Basically, we have too much bad fire, too little good, and too much combustion overall. Even as we ratchet down the burning of fossil fuels, we'll have to ratchet up the burning of living ones. All this is ultimately a fire problem (and a fire narrative) that requires fire-centred solutions (and a fire-themed story). In Australia, especially, it leads to the firestick.

New avatars are appearing. After the 2016 season and some hard-fought debates, Western Australia seems to be reacquiring controlled burning as an official policy. In the interior spinifex region Aboriginal burning is being documented by a new generation of anthropologists as testimony to the resilient power of firestick farming. In the Northern Territory Aboriginal fire practices are co-evolving into modern hybrids; controlled burning is widespread. If upset, even banished, the firestick returns. But the general discourse about 'Australian' fire collapses into the south-east, where bad fires crash into the major concetrations of Australian settlement.

Here the revived firestick centres on the notion of cultural-discourse. The cultural firestick is the traditional Indigenous firestick updated for contemporary times: it is intended as much to restore heritage as to promote flora and fauna. It is seeping into vernacular life much as ritual acknowledgements to 'the traditional custodians of the land' and to 'elders past, present, and emerging'. It promises to disrupt the existing discourse over bushfire, making room for an Indigenous voice.

Yet it will likely be a tweak more than a revolution. Years of experimentation in Arnhem Land and Kakadu National Park show how difficult it can be to synthesise two fire cultures, because fire is a creature of its environment: what works in one biotic, social, economic, legal, and political context may not transfer to another. The Indigenous firestick succeeded in part because it was free to roam across seasons and over large areas, and with accommodations learned over millennia. None of those options will reappear without deep negotiations. How the firestick works in politics is not how it works in country, and vice versa.

Yet cultural burning is a striking re-emergence. More than

another expression of the decolonising trend that also dissolved forestry bureaus, it proposes alternative ways and purposes for burning, more sensitive to the nuances of land, less bound by rules of liability law and property ownership. It allows another firestick to kindle from the flames that all Australians share. It reminds us that the firestick is not just another tool in the tool shed, but part of a conversation between people and the places they inhabit.

Another novel firestick exists at present only as an echo. With regard to landscape fire, Australia and America have formed over the past century a historic fugue, each watching the other for lessons. So what has happened in America since 2005 is also relevant. Basically, bad fires have gotten worse, and good fires have failed to scale up to the dimensions required, and of course no political action at a national level has lessened the contributions of fossil fuels. Yet those on the ground are devising ways to work with flame.

The 'fire revolution' that boiled over during the 1960s and 1970s identified 'fire by prescription' as a foundational strategy. Prescribed fire — with the drip torch as America's firestick — would be the pivot. In the decades that followed, the strategy flourished in the south-east; Florida alone burns a million hectares annually. But the strategy stumbled in the American West, aggravated by several spectacular failures (one of which burned into the Los Alamos National Laboratory, with its cache of plutonium, and racked up costs of $660 million). Prescribed fire as a set piece has become trickier, more expensive, and less agile, burdened with an ever-expanding checklist of pre-burn requirements.

In 2009 an official reinterpretation of federal fire policy opened options for fire officers to reimagine wildfires from problems into opportunities. Fires of any cause, in any setting, for a plethora of mission-appropriate reasons could be 'managed' rather than simply attacked. This has inspired a strategy of point protection for high-value assets such as communities, but has otherwise expanded the domain available for burning out. Fire officers are drawing large boxes, with perimeters set by existing barriers, and then they are methodically firing out their interiors. Done well, the burnouts resemble prescribed fires conducted under urgent but not emergency conditions. They can even be planned.

The upshot is a hybrid: half suppression, half prescribed burn. Box-and-burn is a trade-off. It lacks the rigour and appearance of control promised by traditional prescribed fire, but it allows for more good fire than a program of traditional burning could engender. It's an administrative mash-up that permits fire managers to work with fire. It hasn't yet come at a meaningful scale to California — always an exception (as the Victorian mountains are). But it may find its way to that corroboree of firesticks that have gathered in Australia.

Australia is a continent of fire, of many fires, of endless varieties of fire regimes. With human colonisation it has become equally a land of firesticks. Fire will happen, with or without people. But what people do with fire speaks volumes about how they live on the land, and what they do to reconcile those firesticks says much about how they live with each other. In the end, they all share a combustible landscape on which they are the keystone species for fire.

Environmental critics have sounded alarms that our future is so strange and so dire that we have no narrative by which to bind that future to the past, and no analogue with which to guide our understanding. I disagree. We have a marvellous narrative, the saga of humanity and fire; and we have an apt analogue with the Pyrocene. However much our present condition seems unmoored from anything before it, our moment is historically constructed. It looks the way it does because events have evolved as they did, and it would be wise before intervening to understand the particulars. The present scene did not spring from first principles but from a long, gnarly, unpredictable evolution. The search for a usable past is the flip side to the search for a usable future.

In his canonical report, Judge Leonard Stretton famously intoned the calamity of Black Friday. Yet he spoke in the passive voice. The fires were lit by the hand of man. Villages were burned. Forests were savaged. His judgement is the record of someone reporting what has happened to others — and others not, it seems, able to act for themselves. Above all, people were surprised to know what was possible. 'They had not lived long enough.' After Black Saturday and the Forever fires, surely Australians have lived on their land long enough.

This extended essay is a study in historical context, and it is itself the product of a particular historical moment. Its original thrust was to conclude with the after-tremors of the 2003 bushfire season and the siting of those fires within the chronicle of the firestick. The text was also the outcome of a particular literary strategy to convey that theme. It had an integral design, not exactly the equivalent of a Sydney Harbour Bridge, but far from a pontoon of planks. Its argument led to a climax in the Alpine fires and the political controversies that glowed in their ashes.

To address the sentinel bushfires of this century would require another chapter or two. The book's theme could accommodate further iterations of the firestick — that, in fact, would support its argument. The architecture of the book could not; the new chapters would stumble on as anti-climaxes. The Pyrocene epoch that the 2009 and 2019–20 seasons have unblinkingly announced will demand its own narrative, which will certainly have to wait for a post-mortem on the Forever fires.

This new edition keeps its historical pivot in the 2003 season, lightly updated for more recent events. It keeps, also, the unrevised hope that the comments of a 'scholar on fire' — a friendly critic, although an outsider — might contribute to the conversation that Australians must have among themselves.

PROLOGUE

Earth's flaming front

The first serious shocks came in the 1980s: the 1983 Ash Wednesday holocaust in Australia; the 1982–83 blowout in Borneo; the lurid arc of fire gnawing at the Amazon; and the 1987 and 1988 seasons in the U.S., during which almost half of America's iconic Yellowstone National Park burned, day after day, broadcast on TV like a soap opera. The fires sensitised the media that something was happening, and that the rolling volleys of fire might signal a story deeper than a random shark attack or a Mount St Helens eruption.

Then a drumbeat of conflagrations began in 1993, when fires ringed media-saturated Los Angeles. The 1994 season shocked America; the 1994–95 season stunned Australia, as fires ripped around Sydney and its Blue Mountain exurbs. The fires then mustered and stampeded on a two-year cycle in the U.S., providentially timed with national elections: 1996, 1998, 2000, 2002. They roared into Victoria in 1997, slammed Sydney in 2001–02, and then overran all expectations and incinerated any pretence of control when they swept over the alpine regions and sacked Canberra suburbs in 2003.

A few years earlier, as wildfires lit up the Pacific's Ring of Fire, from the Russian Far East to Mexico, from Borneo to Brazil, the World Wildlife Fund (now the World Wide Fund for Nature) had anointed 1997–98 as the Year the Earth Burned. The cascade of unconfined combustion spilled over Mediterranean Europe, Canada, and Mongolia, and splashed over rainforest and heath. The WWF was wrong; their definitive year was only a midpoint, or worse. There seemed to be no end to the flames, no explanation for their rampage, no meaning to what might appear as ecology's version of an emergent plague.

The reality was different. Open burning on the Earth was, paradoxically, contracting, not expanding. Rather than being an alien visitation, big fires were becoming normal, mirroring, for nature's economy, the boom-and-bust of rekindled stock markets. In Australia, bushfires joined droughts and floods in a macabre calendar of near-predictable crises, linked to the rhythms of the El Nino-Southern Oscillation, exurban settlements, and global industrialisation. And far from rending Australia, its bushfires offered a measure of continuity, bonding its ancient human inhabitants to one another and the land.

Amid this panorama, Australia was sometimes commonplace and sometimes singular, but always indispensable. Australia has more formal art — paintings, poems — about bushfires than any other society. It is one of the Big Five in fire science. Hectare for hectare, few continents, notably Africa, boast more burning; but Australia has by far the more damaging fires. Since European settlement, it has filled up the week with named conflagrations; Red Tuesday, Ash Wednesday, Black Thursday, Friday, Saturday, Sunday, and Monday. It is starting to double back on its days, and has begun naming its holocausts for holidays (Black Christmas) and locales (Alpine fires). But all this is not what makes Australia fire's lucky country. That honour comes because, uniquely among developed nations, Australia kept a tradition of controlled burning.

The practice has many forms and goes by many names: prescribed burning, hazard-reduction or fuel-reduction burning, controlled burning, cultural burning, or simply burning off. It is, in fact, part of a much longer history, older than our species, by which hominins have used fire to make their world more habitable. Peoples everywhere, over all time, have burned. They have exploited fire to catalyse or interact with virtually every other technology and everything they do on the land; through their control over combustion, they have sought to shape their surroundings. We hold a species monopoly over the manipulation of fire, and we are unlikely ever to surrender willingly either the firestick or our hegemony over it.

The capture of flame by early hominins thus marks one of the great divides in earthly history and provides one of the prime indices for our character as an ecological agent. But another historic horizon, almost as momentous, has been the magnification of that firepower by substituting fossil biomass for surface biomass as a primary fuel. This transition is, for fire, the meaning of industrialisation. When nations industrialise, they extract their combustibles from the geologic past and transfigure and route that emergent fire power through steam, dynamos, internal combustion engines, the doubly transmuted fire of electricity, and the like. By both substitution and suppression they remove open burning from the land. The consequences can cascade widely because fire's removal may be as much a shock as fire's presence. Everywhere, over and again, the process has repeated itself in what might be termed industrialisation's pyric transition. Everywhere, except in Australia.

One interpretation is that Australia is a laggard — distant, retarded, perhaps slovenly. The application of force to the removal of fire is something that Western civilisation early declared a value, and Australia has simply not stepped up to the task. Its recurring bushfires are a sign of the still-sluggish process of civilising a refractory continent. If its bushfires are a symbol of the Australian environment's implacable differentness — combustion's equivalent to the platypus and wombat — they are also a stigma of Australia's social indifference, like casual burning in Brazil or Zimbabwe.

The other interpretation is that Australia got it right, and that the nations which Australia's elites strove to emulate wrongly eliminated an ecological presence whose absence is costing them dearly. Certainly this is the case with the United States, whose fire officers have for decades looked covetously at the capacity of their Australian colleagues to put fire on the land. For 30 years the recognition has been widespread within the American fire community that fire's attempted exclusion was a mistake; and the appreciation has grown that the fundamental error was not that fire agencies suppressed wildfires but that they ceased to light controlled ones.

What makes that debate so intractable, however, is that both sides regard Australia's bushfires as inextricably bound up with questions of identity. What is nominally about flame very quickly becomes a discussion about something else. The practice of Australian fire quickly morphs into the politics of identity; geographic, professional, national. The fissures are many and cross one another, like veins in granite. City v. country; greenies v. farmers, graziers, and loggers; ecologists v. foresters; those who live off the land v. those who visit it; those who believe bushfire is ultimately an expression of a nature beyond human contrivance, and those who believe humanity can, for good or ill, profoundly alter fire's regimes. All perceive the contemporary fire scene as inappropriate; all demand that they be heard; and all recognise that bushfire forces society to choose, though what that choice means, or implies, is often as fluid and intangible as flame itself.

Nowhere is this truer than when discussion touches upon 'hazard-reduction burning,' which can escalate into synecdoche not only for the political debate about fire policy but for the whole trajectory of Australia's environmental history. This process of symbolic refraction is common enough, and it would not matter if the 'discourse' were about novels or architecture or political theory. It does matter when it relates to fire because bushfire obeys its own logic and speaks its own idiom, a grammar of wind, drought, terrain, spark, grass and scrub. It isn't listening to the rhetoric, the research, or the parliamentary resolutions. It doesn't feel our pain. It doesn't care. It just is.

PART I

Firestick fundamentals: a primer

Context is all. Unlike the other elements of the Ancients, fire is not a substance but a reaction. It is what its surroundings make it: it takes its character from its context. One does not, for example, hold fire as one might hold water or earth; one holds the ingredients that combine to sustain it, typically, in the form of a firestick. But how did fire happen, and how did it occur that humanity grasped the firestick, and what does that capacity mean? For an explanation, we must expand our sense of fire's ingredients to include a dash of history.

Nature's fire

The reaction itself is among the most elemental known to life: combustion takes apart what photosynthesis puts together. When it happens within a cell, we call it respiration. When it happens outside, we call it fire. What allows combustion to propagate on the land is also a product of the living world. Marine life pumped the atmosphere full of oxygen; terrestrial life provided hydrocarbons potentially primed for burning. When plants first began colonising continents, in the early Devonian, fire appeared. It has burned ever since.

In its simplest formula, open combustion — fire — occurs when two processes, each with their own logic and cadences, meet. One determines the amount of stuff to burn; the other, whether ignition connects with those combustibles. The fundamentals are simple enough. What matters in any landscape are the rhythms of wetting and drying: conditions must be wet enough to grow fuels, and dry enough to ready them to burn; the wet–dry cycle acts on a biota like a frost–thaw cycle on rock, cracking it and creating fissures into which flame may penetrate. Some places thus experience the conditions for fire according to an annual cadence of seasonal aridity or with decadal drought; others, rarely if at all. Thus normally well watered woodlands burn during periods of drought, while deserts carry fire after periods of deluge, which allow combustible carpets of grasses and forbs to sprout temporarily. Places that are chronically wet or chronically dry don't burn What matters are the deep rhythms that flush and drain.

Even so, burning demands a spark to set it off. Spontaneous combustion is rare, far too infrequent to account for the pervasiveness of fire on Earth. And while ignition sources are several, the only one with sufficient clout and range is lightning. Fire is a two-cycle engine for which, in nature, lightning supplies the spark. Ignition, too, obeys a logic of wet and dry, for what matters is dry lightning, that is, lightning separated from drenching rains. Maps of lightning and lightning-kindled fires overlap poorly: the fires come on the edges of storms, or the margins of thunderstorm regions, or during the first tremulous onset of the rainy season.

Under natural conditions, that is, fire's appearance depends on these compounding cadences, and is lumpy in both space and time. There were (and are) places that experience fire regularly and places that know it sparingly, if at all; there are times rife with flame, and times empty of it. Even with oxygen and plant life at higher levels than today, the Carboniferous era simply buried vast quantities of biomass. While those coal-bearing strata are typically laced with fusain, or fossil charcoal, the capacity to make fuel seemingly exceeded the capacity to combust it.

The coming of the firestick

This dynamic changed abruptly when early hominins captured fire. No longer did the Earth's biota need to rely on lightning's lottery: humans carried flame everywhere, and they made fire a more or less constant presence. What fire did for landscapes, it did for humans through cooking. Cooking boosted our caloric intake, it remade our habitat, it gave us small guts and big heads. Fire entered our genome; we could not survive without it. Like so many other species we adapted to seize its properties, expanding its range to landscapes and ultimately the planet. Unlike others we had some say over where and when it could kindle.[1]

How soon, how far, and with what outcomes the firestick left the hearth are difficult to determine. But fire differs from the otherwise mechanical implements in a number of ways. Like them it can apply concentrated force: one can heat a stone to the point of cracking, warm a cave or illuminate a windbreak, burn down and hollow out a tree with a focused flame. A flame can rest on a candle the same way an axe head can sit on a handle. But unlike stone and bone utensils, fire is also a biotechnology, feeding upon and insinuating itself into the living landscape; it requires constant tending, an early act of domestication; and, most spectacularly, it can propagate. It takes a great deal of effort to chop down a tree; it takes but an instant to set a fire that can burn down a forest. Knowing how to sculpt a landscape by fire required a different level of understanding than knowing how to cook meat or temper a wooden spear over a flame. Yet there is little to reason to doubt that early humanity quickly appreciated fire's power.

Myths on the origins of fire testify to its inestimable worth. The usual mythic narrative has a scruffy humanity, unendowed with any special attributes, weak and wary among the grand animals, reduced to warming food by the sun. They get fire by accident, or stealth, or sometimes by force, and at that instant the world changes. They move from prey to predator; the ecological chain of being is upended; people possess a power unique among creatures, an emblem as identifying as an elephant's trunk or a tiger's stripes. More than anything else the firestick would brand humanity's presence on the land, not only by setting fires but as a catalyst for practically every other action people take. Possession of the firestick became, in truth, a species monopoly, the distinctive emblem of our ecological agency.

What we might call aboriginal fire — the ability to make fire and put it on the land — was both enormously potent and dismayingly feeble. The firestick was a force multiplier, an ecological lever that, suitably positioned, could move continents provided it had a suitable fulcrum, for the power of landscape fire resides precisely in its capacity to propagate: that is both the glory and liability of the firestick, which can only spread spark where nature will allow. Places that already knew fire could be seized and their fire regimes redirected to new purposes; places that underwent routine cycles of wetting and drying but lacked a spark could now burn; but places that remained permanently wet or dry would extinguish whatever flame people applied.

It is no accident that so many myths of fire's origin sequester fire in the wide world, where people must extract it from flint or grasstree. Fire came from nature, and from nature it would derive its force and its fragility. People did not invent fire: they could call it forth. People did not spread fire: they could place it where wind and scrub and slope would allow it to smoulder, flash, creep, and ramble through the countryside. They could grasp whole landscapes in places already combustible; they could tinker at the margins of sites susceptible of burning, with effects that over centuries or millennia could reconfigure fungible biotas; they could set off chains of ecological reactions that, by removing or rearranging species, could affect still others. But they could not burn through snow or under rain, could not have flame race against an opposing wind, could not make wet woods dry or forage for continuous combustibles in stony deserts. The firestick could coax, it could not conjure. It remained a creature of context.

What people did introduce, however, was a radically new context, culture. They did not burn according to instinct; they burned by choice. While they came genetically equipped to manipulate fire — they could fashion a firestick and carry it — they did not come into the world intrinsically knowing how to use the device. The software had to be learned. For good or ill, the firestick thus entered a moral universe scripted by human values, beliefs, understandings and fallacies, skills, and clumsiness. People could rearrange fire's regimes in ways that competed with and defied the putative natural order. And they would judge the outcome according to how the firestick behaved within the imagined world in which they lived. The image of a kindled fire, called forth from flint or wood and cast about the countryside, captures exactly the essence of that mutual relationship.

Firestick Australia

Australia, too, came predisposed for anthropogenic fire. Its long migration after the break-up of Gondwana had tempered its biota to survive drought, disturbance, oft-impauperate soils, and selective extinctions. The Australian ark held a toughened flora and fauna, ready to scavenge and hoard necessities, to respond quickly to flushes of water and nutrients and, at least in places, to accept and even encourage fire. But what served one purpose likely served others as well; Australia's flora boasted suites of adaptations that accommodated suites of stresses, such that a trait that helped survive drought probably also helped to survive the fires that could follow. The biota was salted with species ready to respond to the advent of regular fire or to a shift in fire's regime.

Little of Australia did not exhibit a suitable regimen of wetting and drying. There were annual cycles tied to the tropical monsoon in the north, and to Mediterranean climates in the southwest and southeast, and even temperate climates displayed regular spells of dryness that were especially useful if they came before greenup or after dormancy. Beyond this was a decadal cadence linked to the droughts and deluges of the El Nino–Southern Oscillation, leaching moisture away from normally humid mountain flora or drenching with lush growth an otherwise arid interior. And there were yet deeper rhythms that kept most, though not all, of the continent primed to burn whenever spark could catch tinder. Where ignition flashed at the right times, fires kindled. But it is also likely that much that could, potentially, burn did not, for which one can point to the fickleness of lightning.

This, of course, is exactly the condition that changed once firestick-wielding peoples arrived. No longer was ignition a limiting factor: it was constant, if latent. Its success, in fact, depended on its being used selectively, for wanton burning was a kind of ecological vandalism that could damage rather than enhance. The firestick, that is, had two checks on its application. One was nature, which determined whether its spark could propagate; the other was culture, which sought to guide the hand that held it.

The landscape sculpted by the Aboriginal firestick is largely vanished, along with the pragmatic memory of its quotidian use. Only in a few places does an environmental record remain of what Aboriginal-inspired regimes must have been like, and mostly through scattered historical documents is it possible to gather clues about how, prior to European contact, the firestick worked on the landscape. The reality is that the archives are even worse than such observations suggest. There is, first, scant evidence for Aboriginal fire's First Contact: how the firestick met prelapsarian Australia; how it interacted with oft-dramatic shifts in climate, extinctions among flora and fauna, and other human practices; how, out of this swirl, some sort of sustainable fire regimes emerged. And there are, secondly, difficulties reconstructing Aboriginal practices based on sources from European encounters. The hard evidence is light, and the interpretive burden intense.

Start with the conundrum of first contact, of which two polar versions exist, each untenable. One suggests that Aborigines advanced across the continent behind a line of fire, like that scene in Star Trek's *Wrath of Khan* by which a Genesis device burns over a dead planet, remaking that world 'in favour of its new matrix'. The other proposes that Aborigines — few in number, outfitted with very primitive technologies, unwilling as well as unable to reshape their habitat — did little more than dapple the Dreamtime with billy fires. That either version has any credence bears witness to how laden with metaphysical burdens and overwrought rhetoric the subject has become. Nor is an answer 'somewhere in between'.

The Pleistocene, when humans apparently ventured into what is now Australia, was an epoch of exaggeration — of vast climatic tides, of coastal flooding and migrations, of giant creatures, from huge serpents to waddling land-crocodiles to wombats the size of ponies. So much was going on that it is tricky to tease out what impact the colonising firestick might have had, and so little can be dated with the necessary precision to identify the firestick as a smoking gun. Various lines of evidence argue that Australia had plenty of prelapsarian fire; that it had been on a trajectory of burning for eons; that, in the words of D.M.J.S. Bowman, its Gondwana-relict rain forests were islands amid a sea of flame forests.[2] But some general observations about the impact of the Aboriginal firestick are surely in order.

There is no evidence of any people not applying and withholding fire in order to make their land more habitable. Fire was the most powerful of Aboriginal technologies: they used it. But how, and to what effect? Where the land already had flame, they would ensure flame remained, but likely shifted the timing and extent of burning. Burning would be more regular, more extensive, and more textured; the regime would change. Where the land had the conditions for fire but no regular ignition, Aboriginal inhabitants would provide that spark. Where the land was hostile to fire, it would remain so. The firestick could kindle but not slash: it could not ready a fire-intolerant site for flame, and could not force fire through country where fuels and wind resisted. Without an ecological fulcrum, it was a stick, not a lever.

Fire is the most interactive of technologies, and more interesting perhaps is the question of how fire and the extinction of Australia's Pleistocene megafauna compounded one with another. In some respects the query only magnifies the uncertainties, balancing one unknown against another. But the disappearance of megafauna seem to correlate with Aboriginal colonisers rather than with climate. The addition of a new top-of-the-food-chain creature can tip quickly the always precarious balance between prey and predator, and in this instance the newcomer could also restructure the landscape itself through a controlled contagion of combustion. Once begun, each reaction could set off another such that ecosystems could collapse. Recent research proposes this may have happened.

The Aboriginal colonisation is believed to have occurred between 65,000 and 45,000 years ago, coinciding with the dated extinction of most large animals between 50,000 and 45,000 years ago. Revealingly, 'the disappeared' were all browsers; those megafauna that survived, like the red and grey kangaroos, were grazers. This suggests a shift from shrubs to grasses (and within the grasses from C4 to less palatable C3 varieties), a staggered break recorded in dietary differences which favoured species that foraged omnivorously or on grasses, and selected against specialised browsers. Thus, the indiscriminant emu survived while a giant flightless rival, the browsing Genyornis, did not. As incredible as it sounds, the evidence — a 'stronger case than ever', as one scientist puts it — is that 'the arrival of humans had a larger impact than the last glacial cycle on ecological change in Australia'.[3]

The story does not end there, however. The removal of browse and large browsers, nudging the land into more grasses, made more fuel readily available which could feed more fire. In this way, bit by bit, the firestick — assisted by the spear, by species emboldened under the new regime, and by favourable climate — could nudge the landscape into new patterns. Ethnobotanists have often recorded the astonishing variety of plants used by nominally primitive tribes, with the implication that they are both highly knowledgeable and tightly adapted to their setting. Unable to alter the world, they have learned their circumstances and accommodated themselves to it. An alternative reading is that they have, over the *long durée*, removed those species they couldn't use and promoted those they could. While such studies typically relate to slash-and-burn cultivators, there is little reason to doubt that the same logic applies to hunting, fishing, and foraging economies as well. Especially if the circumstances favour fire — and most of pre-contact Australia was either burning or primed to burn — the prospects for wholesale transformation are excellent. In fact, it is difficult to imagine how such a reformation could not have happened.

Few things are less secure than today's scientific revelation: tomorrow will bring another. Moreover, each new data point is immediately dispatched to the ideological front. The issue is not that Aboriginal colonisers brought fire where it did not exist: the question is in what ways they altered the patterns of burning, which they surely did, by means both direct and indirect. If the capacity to kindle fires resided in a new species of insect or bird or, especially, a predatory megafauna that had arrived in Australia 50,000 years ago, no one would argue that nothing had changed in the continent's fire geography. They would, rather, search out the dynamics of how the newcomer nudged and jolted the landscapes it encountered and how the rest of the biota accommodated its presence.

The complication, of course, is that the newcomer was human, which means that the firestick is not merely a tool but a symbol, and that it carries not only a chemical reaction of heat, hydrocarbons, and oxygen but a cultural one of meaning, choice, and judgement. It illuminates all the ambiguities of what it means to be human, of who we are and how we should behave. Whatever the ways Aboriginal contact remade natural Australia, its more profound impact was to transform that landscape into a moral universe in which the flames become an unstable and distorting pyric mirror. The story is a cipher, coded for understanding by today's disputants, with the colonising Aborigine as Future Eater, Noble Ecologist or, perhaps most simply, the Earth's Fire Monopolist. (It would be interesting to know not only what those first colonisers did but how they understood what they were doing, and whether some of them objected to spreading fire where nature, in its evolutionary wisdom, had not put it.) But cause, consequence, catalyst — the burden of understanding all collapses into the meaning of the firestick.

For decades, Europeans doubted that Australia's Aborigines had the capacity to start fires because, instead of flint and tinder, or firebow, they invariably carried their match in the form of a smoking firestick. They carried fire on their treks, they kept it in their canoes, they kindled it whenever and wherever they halted. The earliest picture of Aborigines shows a family returning from fishing, with a young boy holding a firestick. The Aborigine and the firestick were inseparable — symbiosis may not be too strong a term to describe their interdependence. The people might change, and the particulars of the firestick alter, but the bond between humanity and fire would endure, and what affected one would affect the other. In brief, since it first arrived, the firestick has never left the Australian scene, and has never wandered far from controversies over how Australians should live on their land.

Firestick history: a synopsis

Firestick farming: Aboriginal Australia

Evidence is a bit surer from the time of European contact, and in truth constitutes some of the best historical documentation available anywhere about firestick practices. One reason is that Encounter coincided with Enlightenment, which meant that most of the educated elite, including military officers, had an interest in natural history and that the culture's general secularising trends had replaced missionaries with naturalists. Those recording observations were interested in flora, fauna, and native peoples in ways different from, say, Spanish missionaries to Mexico and South America who paid close attention to political alliances, native religions, plants useful for food or medicine, and the location of mines, and little else. By the time Enlightenment explorers began to record systematically the natural landscape, indigenous societies (and their sustaining biotas) had undergone a demographic collapse and were a shambles, only partly reconstituted. This was not the case with Australia. The records for fire use are relatively abundant.

This does not imply that a reconstruction of Aboriginal practices and fire regimes is simple. It isn't. But with the coming of the firestick new rhythms of burning appeared and became endemic.

Not everywhere, and not everywhere with the same consequences — how could it be otherwise? — but over time, and Aborigines had tens of millennia at their disposal, anthropogenic fire became as elemental as sun and rain, gums and possums. The reasons for burning were many: to clear and clean, to promote new growth favoured by prey animals, to hunt by fire drives, to expose burrows, to signal, to claim ownership; a full roster is almost endless. So basic is fire that it accompanied virtually every activity, and it is through that interaction, not simply burning itself, that fire began moulding the scene. A burn could be as precise as a campfire in a tree hollow that smoked out possums, or as vast as a fire hunt in spinifex that could ramble through the countryside. But whether big or small, whether set annually or only when fuel and drought allowed, whether routine or opportunistic, such fires created a cultural, if not domesticated, landscape. As Sylvia Hallam reported, 'The land the Europeans found was not as God made it; it was as the Aborigines made it.'[4]

There was method to the flames. Fires tracked people: they gathered, especially, along corridors of travel and at sites dedicated for special hunting and harvesting. Call the first lines of fire, and the second, fields of fire. Together they inscribed an elastic matrix within which lightning fire had to coexist. This is the basic geography of burning. But there was also a calendar of burning, a kind of fire songline tied to an annual round of hunting and harvesting, fishing and foraging. The season would begin typically with the earliest cured sites; piddling, smouldering fires, gasping for fuel amid still-green landscapes. Then, as more of the country dried, the fires would multiply and spread. By the time the rainy season, and lightning, would appear, those areas Aborigines wanted burned were already burned, and those they wanted protected were shielded from any fires started by random lightning, or for that matter, by accident. Tame fire confined feral fire. Fires of choice replaced fires of chance. The whole process, as intricate or coarse as conditions permitted, represented a kind of alternative cultivation, what Rhys Jones famously labelled 'firestick farming'.[5]

The image is attractive, plausible, and easy to overstate. The limitation, as always, was that the ability to start a fire did not guarantee its spread. Thousands of years of calculated burning, along with the extinction of megafauna and other events, could push the land into new arrangements, but could not trump the fundamentals of terrain and climate. Thus the same practices could yield very different patterns in the wet-dry tropics, the Mediterranean climates of southern Australia, the spinifex deserts of the interior, the cool-temperate forests of Tasmania. There was a uniformity of practices, but a diversity of outcomes; the Aborigine could control the timing of a fire but not the conditions that promoted or prevented its propagation. Firestick farming relied on a nimble capacity to match flame and fuel quickly.

A more profound discovery concerns the Aboriginal sense of burning as 'cleaning' the landscape. The cleansing fire was a means to make land habitable and to display, as it were, that one had done one's biotic duty. 'Corrective burning', as H.T. Lewis observed, was instituted 'irrespective of the time of year since, in the view of informants, further delays can only make a bad situation worse'. Proper burning signified ownership, a duty of care to the country. Aborigines, Lewis continued, 'sometimes give the impression of having an almost manic compulsion about re-establishing fires in neglected environments'. The firestick was an instrument of judgement. Its fulcrum was moral as much as biological. Still, not all fires were deliberate or showed either good judgement or good behaviour. One should not overlook simple fire littering, the ubiquitous dispersion of sparks from caprice or carelessness, or the unforeseen collusion of wind and ember which could send a controlled flame into a conflagration. The reach of the firestick could far exceed the grasp of those who held it.

Always there were limits. There were places and times to burn, and not to burn, and these were set by the grand ebb and flow of wet and dry, the local peculiarities of wind and terrain, the threat of wildfire from lightning or accident, and an understanding of need. Flame could only propagate where natural conditions permitted, but the firestick could kindle only where culture willed.

The pattern of fire on the land is generally expressed as fire's regime. When foresters dominated the subject, they interpreted fire as doing pretty much what foresters do. Thus, fire 'plants', fire 'prunes', fire 'harvests', and it does so in a 'cycle' that mimics the rotational rhythm of logging. This image is altogether too mechanical. Rather, the fire regime is a statistical composite. Fires occur within a regime much as storms do under a given climate; a given regime can accommodate a variety of fires beating to a variety of tempos. There is always the exceptional storm, a cyclone or a freakish tempest, quite outside the norm but still within the gross parameters of the climate. So, too, with fires. And just as the monsoonal tropics of the Top End differ from cool-temperate Tasmania, so likewise their fire regimes, in the end, look quite distinct.

What matters is that it is fire's regime to which the flora and fauna adapt. It is meaningless to say that such and such a species is adapted to fire: one might as well say it is adapted to water. What matters is the patterning of precipitation — does the rain come with equanimity, month by month, or is it restricted to a long season, or does it fall in spasms? A plant adapted to one rain regime would find itself maladapted if the pattern changed. It is much the same with fire, and it is here that the firestick wrought its profound impact. Outfitted with firesticks and hunting sticks, Aboriginal peoples could, in places like Australia, reshape those regimes, and once established, those regimes adjudicated the conditions to which so much of the biota had to accommodate itself. A radical change in regime — the removal of the firestick, for example — could destabilise the country as surely as the removal of a rainy season.

But the interpretive meaning is equally cultural. The documentation is inevitably inadequate to the burden we wish to place on it. It is possible to dispute the pervasiveness of Aboriginal burning, and its ecological impact and, especially, the moral significance it bequeaths to the present day, for it is the latter that tends to drive the former, the answers that set the questions. If the Aboriginal firestick was, in fact, integral to Dreamtime Australia, there is a case to keep it somehow on the land if one wishes to preserve other elements of that indigenous country. If the firestick was irrelevant, or perhaps malicious, the conclusion follows that one should prohibit it: for what the firestick itself did not damage, it made possible for other practices to do.

Behind these arguments lie deeper polarities — rhetorical straw men — that are quickly sketched, and dismissed. One group implies that because people have always done what they thought would improve their condition, whatever they wish to do is sanctioned. Those who inhabit the land today are merely doing, for our times, what Aborigines did in theirs. Another group claims implicitly that the Aborigine was incompetent to make such changes, and if he did, the fundamental truth remains, that everything people do worsens the environment and the wisest strategy is to do nothing. At this point we are immersed in a contemporary version of the Dreamtime, a moral universe of symbols and archetypes and narrative anecdotes as songlines that intend to tell us where and when we might carry our contemporary firesticks.

Yet, while oft furious, the singular focus on the firestick is misplaced. The firestick never acted alone. It went where the mind that directed the hand placed it; it always interbred with other practices, for which it was the great enabler; its significance lodged within a larger economy and belief system. That past is gone. But some elements, good and bad, persist, and others are worth hauling into the future. If there is no going back, the issue remains what to carry with us as we move on. It should surprise no one that, whenever the firestick has been put down, someone else has picked it up.

Firestick farming: European Australia

At first the firestick fitted uneasily in the hand of European settlers; but, over time, rather quickly in truth, they became inseparable. But now the conditions — the context that makes fire what it is — had also changed. Europeans, conveying their agricultural heritage with them, brought devices to leverage fire's power. They carried implements to modify fuel as the firestick modified ignition. They brought hoof and axe; livestock to chew and trample, and axes to slash and ringbark. Now people could create fuel as they did spark. The axe does for the firestick what a woomera does for a spear: it adds impact.

For fire history, this capacity to convert landscape into fuels is the meaning of agriculture. People begin to slash, drain, grow, loose livestock, or otherwise alter the vegetation in ways that make it possible to combust what, by nature alone, fire could not touch. Typically, more area burns, and old areas burn in new ways. Chronically wet areas, for example, or woods that burn only during appalling droughts, could now be felled and minced into combustibles that could dry and burn out of season. Fire could get into areas it couldn't before, could burn at times it couldn't previously, could burn under different rhythms, could catalyse different practices, and could associate with novel plants and animals. Farming and herding were powerful forcing mechanisms: they could, within limits, defy natural barriers and compel burning where it would otherwise snuff out. All this became possible because people could augment the stuff available for combustion.

The power of cultivated fire depended on the capacity of people, working with the land, to create fuel.

Outside of floodplains, where water did what fire did elsewhere (where floods could purge and promote), agriculture needed fire somewhere in the cycle of cultivation. That required ample fuel, which encouraged new clearings and, within old ones, the practice of fallowing. The first brought fresh lands into the system, while the second renewed stale lands. Another round of burning required another stockpile of fuel, and in an agricultural system, the way to stockpile combustibles was to grow them. Directly or indirectly, agricultural colonists began rearranging the vegetative cover in ways that distorted the old geography of fire regimes. No longer was fire only a catalyst for hunting and foraging but for farming, herding, logging, and town-building. No longer was the firestick restricted to times and places set by dry spells and wind. If its domain was not yet unlimited, it could reach almost anywhere people could coax out enough hydrocarbons from the land.

In Australia, this reformation was the contribution of British colonists. Fire's regimes changed, sometimes favourably and benignly, often with unexpected consequences and outright violence. The process was two-fold. It involved the introduction of new fire patterns, associated with a restructuring of fuels, but it also meant the extinction of Aboriginal practices. Each could unhinge an ecosystem.

The firestick — in some form — typically remained. Since

settlers often relied on Aboriginal labour, the firestick still moved over the country. But more and more it passed into the hands of the colonisers, even if they disguised the flame in strike-a-light pouches and Vesta matches. They became as dependent (or addicted) to their firestick as had their indigenous predecessors. Soon, not a few observers scorned what they considered the casual, promiscuous burning habits of the bushman who lit billy fires whenever he halted in his wanderings, an echo of Aboriginal habits in which every walkabout became a dotted line of campfires, rekindled firesticks, and embers scattered like dandelion seeds.

Writing in 1927, E.H.F. Swain traced the 'firestick habit' from Aborigine to the white pioneer, the squatter who discovered that 'by using the aboriginal method' he could improve pasture and the selector who found fire 'to be the settler's blessing'. Since then Australia 'has been burned and grazed, and burned and grazed as it never was burned and grazed before'. The rural population was by now 'so enamoured of the use of fire that wherever he now goes, and in the periods of highest fire hazard ... the bushman blithely distributes his matches, and the schoolboy is learning to follow in father's footsteps'. Another self-styled bushman informed a royal commission with a verbal shrug that 'the whole of the Australian race have a weakness for burning'"[6]

What had radically altered was not ignition but fuel. There was no simple syllogism at work: some places lost available fuels, some gained, all underwent change. The various insults of colonisation such as trampling, felling, and unleashed exotic fauna, like boulders tipped over a ridge, triggered an ecological rockslide. The effects varied, and often ran counter to one another.

The extinction of Aboriginal fire practices could encourage an excess of combustibles, no longer cropped off by routine flame while, beneath swarms of sheep and rabbits, the grasses vanished, and took flame with them, such that less palatable woody scrub overgrew the scene. Elsewhere settlers avidly hacked forests into fuel wood, fencing, and fallow, culling choice parts for saw timber and discarding the residue as slash, and elsewhere simply ring-barking and abandoning to drought. Like a vast kaleidoscope, the Australian biota was cranked and the clattering shards fell into new patterns.

Lines of fire, fields of fire — the old logic held, such that fires tracked roads and runs, and rose and fell on cultivated fields. Once stabilised, the agrarian landscape imposed its own rhythm on flame and built in new buffers for burning. Fires burned stubble and fallow; land-clearing pyres had their unwritten rules for when and how to burn; squatters fought bushfires on lowland paddocks thin with pasture, and set fires in wooded hills after summer grazing. Compared to Aboriginal occupation, there was more fire in some places, less in others. The big change was that most of the country was unsettled. Much of Australia had effectively become fallow, like a field gone to weed. Once tended by the firestick, it now lay sullen, ready for a spark.

Colonisers spilled over the continent — the place was awash with swagmen of all stripes. It was not simply that the newcomers, once freed from their close-watched archipelago of prisons, were mobile; so, famously, had been the Aborigine. It was that their firestick did not follow 'the law' that had, over thousands of years, matched flame and fuel on a land notorious for its variable rains and droughts. These fires, like rabbits, ran wild, and the shock wave of settlement sparked, with eerie fidelity, a chronicle of conflagrations. The 1851 Black Thursday fires coincided with the Victoria gold rush. The 1898 Red Tuesday fires gorged on broken and ringbarked gums in the Otways. The 1939 Black Friday fires, loosed by unrestrained burners, fed on the carrion of drought-blasted and slashed woods and, likely, the scrub-thickening legacy of grazing throughout the Alpine ranges.

Such fires got recorded because they savaged a literate civilisation. They did not announce the advent of large fires, which are surely ancient, but a change in their character. They proclaimed a new cycle of conflagrations in bitter valence with the disrupted landscapes. They were the fires of regime change, and the after-shocks of the ecological insurgency that violent change inspired — not a new, more or less stable regime themselves. They were mythic fires, collectively a kind of Götterdämmerung that shuttered the Dreamtime and announced a new Australian Asgard from their ashes. Why? How? Not because of their size but their ferocity, the combination of size, savagery, and tempo, along with the menagerie of plants and animals disgorged from the colonising ark. These were likely fires to which the indigenous biota was no longer adapted, and out of which, because of the aggressive presence of the introduced creatures and their campfollowers, the old world could not re-establish itself.

Revealingly, these new landscapes were typically as fireprone as the old, for drought and dry winds persisted, and Australia quickly selected against species unable to weather aridity and flame. In principle, once the frontier had passed, the scene could calm, reconstitute itself, and replace feral fire with tame fire, and there were ample places where this occurred. The overall thrust of European settlement brought, in the end, a vast reduction in burning. But the ferocity of Australia's fireproneness meant that such settlement, to snuff out fire, had to be more meticulous than Australia could easily manage, and that under such conditions a fire-intolerant flora could never replace a fire-intoxicated one. This left undomesticated landscapes as points of fire-infection from which wild fire could spread. There would always be such places, and so long as they endured, so would bad burns.

But what may be most striking was the rehabilitation of the firestick. The kaleidoscope of colonisation turned on it, as a vault door on a bearing. Controlled fire became a treatment of choice, not simply because agriculture was necessarily a fire–fallow economy and not solely along the ecological frontier but as an enduring means to render the land habitable and contain wild fire. In an odd, serendipitous way European Australia was reassembling itself as a firestick-sculpted landscape. The classic literature of bush settlement — the short stories of Henry Lawson or Steele Rudd, for example — are chock-full of offhand references to billy fires, bush fires, burning off; fires of all and every ilk. They simply occupy the country, like cockatoos and wallabies. Competence in the face of bad fire, skill in the execution of good, these were part of the lore of survival, as much as building bark humpies or shearing.

Over the years the process became more rather than less entrenched. Australia, it seemed, would burn. The issue was how, and who would hold the firestick. The imperial forces that pushed settlement had also spawned a counterforce that saw fire less as an inevitable if awkward ally than as unregenerate and ill-tempered enemy. Colonisation met conservation. The fight between them would, in Australia, become a fight over the firestick.

Beyond the Black Stump

European colonisation by agricultural settlers had an accomplice in environmental havoc for, increasingly, imperialism also meant industrialism. The ship's fires that, for the First Fleet, had cooked meals, were more and more contained in coal-fired engines that used steam to drive the ships. What makes such technological reforms relevant is that something similar happened with the land.

As much as the axe ratchetted up humanity's firepower, the firestick still had its limits. It could burn only as much as living fuels allowed, and there were severe constraints to what people could coax or coerce from the soil. You could exceed those limits, but only for a while, with each fiery overdraft forcibly deducted from the accumulated biotic capital, not just the interest. Such burns were extractive fires, a kind of fire mining; and under its demands nature's economy of combustion would spiral down, or simply crash. The first-contact fires of colonising — feasting on heaps of slashed woods — were a fire equivalent of placer mining, a quick flush of wealth that could not be sustained. The fallow that succeeded was invariably less robust, and in the most desperate circumstances left farmers working the agricultural equivalent of tailings. If people wanted more firepower, they would have to find another repository of fuel. They could not pioneer forever, endlessly fossicking for new caches of combustibles. The critical innovation came by discovering ancient landscapes, that is, geologically cached fossil biomass.

Suddenly combustibles became, for practical purposes, unbounded. Humanity commenced to sublimate its firesticks into steam, and to stuff its torches into mechanical combustion chambers. Its burning was now indirect, not with flame on the land but with applied heat and horsepower through engines that chewed up woods, with roads that defined new corridors of travel, with a rearranged geography and recalibrated market of what in the land was valued and what was waste. By means of fertilisers and biocides smelted from fossil biomass, industrialisation created a kind of fossil fallow, abstracting fire even from its ancient alliance with agriculture, moving fire's presence beyond the black stump.

Initially, industrialisation's machines added impact to the axe; rails carried it deeper into the countryside, sparks belched from smokestacks, and the old bad habits became more powerful. Eventually, however, internal combustion promised to replace the firestick and to check even nature's fires. A new species of controlled burning could do without recourse to open flame those tasks previously endowed in the firestick, and by rapid transport, power pumps, and other mechanical devices the new order could snuff out free-burning fire of all varieties. It changed, as it were, the operating system for nature's economy of combustion. By technological substitution it could replace the firestick; and by outright suppression it could drive it from the land.

This radical reformation, still underway, defines the geography of fire on Earth today. The planet is rapidly fissuring into two great combustion realms, one that burns surface biomass and the other, fossil biomass. There is scant overlap between them. How that transition occurs, and whether it displays predictable patterns, has been little studied. The simplest analogy is that the population of fires obeys a similar trend to that for people. There is, initially, an explosion of burning; the old practices persist, while new sources of ignition and fuels scatter about the countryside. But gradually, over the course of 50–60 years, by a process of substitution and suppression, the new mechanical firesticks replace the old, and flame recedes from the landscape. The population of fires drops below replacement values. Countries go from fire-flushed landscapes to fire-vanished, or fire-starved, ones.

Thus industrialising countries typically experience a surplus of burning, often abusive, while industrialised countries suffer from a deficit of benign burns. What complicates the story for many settler societies like Australia is that European imperialism became the vector for industrialisation. The process of agricultural settlement had barely begun when the industrial revolution overtook it; colonisation by the axe quickly coupled with colonisation by steam. The shock of those two processes, with wrecked landscapes dappling the European imperium and spreading like a contagion, alarmed thoughtful observers and underwrote a doctrine of state-sponsored conservation. The most visible emblem of that ecological havoc was an apparent plague of fires.

PART II

Fire conservancy

Conservation affected Australia's elites as it did those elsewhere throughout a Greater Europe and its neo-European progeny; all found in the contagion of fire a practice to condemn, an enemy to fight, and a symbol of reckless and extravagant waste. The wreckage of settlement seemed especially egregious where it destroyed forests. But for every tree killed by axe or trampled by hoof, common wisdom held that ten died by fire. Indeed virtually every abuse that alarmed critics seemed to have the firestick as an enabler, and every attempt at colonisation sparked a pestilence of flame. The solution, a stopgap, was to break the causal chain by which people entered, upset, and burned.

The preferred strategy was to fence off critical places from settlement altogether. Primarily, this meant forest reserves, which in colonial settings could become extensive. While there were some precedents for mountain reserves on tropical islands (e.g., the Caribbean and Mauritius), the surge followed imperial Europe as it moved inland from its coastal enclaves in the late 18th century. For Britain the breakthrough came in India, and for France, in Algeria. To administer the reserves officials turned to foresters, a self-declared profession, a global cadre of engineers akin to those who built railways, opened mines, and erected dams and cantonments around the world. Not having its own foresters (for that matter, not having forests worth the name), Britain recruited the founders of its Indian Forest Department from Germany. Dietrich Brandis, Wilhelm Schlich, Berthold Ribbentrop — these were the founders of colonial forestry, legendary figures in the global saga of conservation, all destined for knighthood.

Their ideal was the forestry of central Europe; the silviculture of Germany and the bonding of forestry to state interests characteristic of *dirigiste* France. It was a misplaced model in two respects. Temperate Europe had no natural basis for fire; what fire existed did so because people put it there as a requirement of fire-fallow agriculture. Fire's regimes, that is, were the manifestation of a social order that both moulded fuels and applied flame. In an intensely cultivated landscape fire had standing akin to the hearth in a house or the forge in a factory. It had a place as a tool, and it was task of authorities to keep it in its place.

The experiment had mixed results. Where the indigenous people remained, as in India, Cape Colony, Cyprus, Algeria, or Senegal, officials were unable to eliminate anthropogenic burning. Over and again, colonial foresters in fire-prone lands found themselves in bitter, chronic firefights with locals who regarded restrictions on their use of fire as among the most oppressive of the many arbitrary statutes enacted in the name of Western rationalism. Whether or not they were allowed physical access to the woods, without fire they had no biological access; they could not extract from the land the forage, the medicinal plants, the honey, the game, the fruits and nuts they required. Often a fire insurgency sprang up, with locals burning under extreme conditions, which only reinforced the belief that fires, all fires, sprang from human malevolence and had to go.

But where the indigenous peoples were more or less removed from the land, control was possible, and in principle this meant fire exclusion according to the European ideal. It is no accident that those countries that succeeded in gazetting great swathes of relatively uninhabited public (Crown) land are precisely those with fire problems similar to Australia. Canada, Russia, and the U.S. all underwent a comparable transformation. For each, a cosmopolitan corps of foresters sought to beat down fire with a mix of bureaucratic authority, applied science, and zealotry.

'Fire conservancy,' as they termed it, succeeded, at least during the early decades. Forest guards could restrict human access and burning, they could attack what fires nature might throw their way, and they effectively occupied those largely deserted lands in the name of fire control. This was what progressive conservation meant: remove the firestick and you disabled many other abusive practices for which catalytic fire was indispensable. For swidden farmers, there would be no slashing if they could not also burn; for pastoralists, no transmutation of timber to forage without fire's ecological alchemy; for hunters and gatherers, no control over habitat, no say over the arrangement of browse and covert, without control over flame; for colonists, no wholesale conversion of landscapes to field and paddock without fire to sweep away the debris of the old and to stimulate fresh growth. This whole swarm of ecologically evil practices would blow away like smoke if only one could seize the firestick from ignorant indigenes and self-centred settlers. In return, timber would become plentiful, catchments would stabilise, and climate would ameliorate and become less prone to moody swings between drought and deluge.

All this made controlled burning anathema. The point of seizing the torch was, ultimately, to extinguish it. When this proved more formidable than anticipated — 'tropical' lands with their dry seasons were especially refractory — officials frequently found themselves exploiting fire as an expedient, particularly if they relied on a fire-familiar native work force. Such burning might be tolerated as a necessary and temporary evil until proper administration could be installed, but rightminded foresters knew their cause would prevail: they had modern science, their fire-wielding antagonists only folk habits. They needed only political will to triumph. Foresters fought bitterly against those they perceived as feckless burners, whom they denounced either as cynical exploiters or superstitious settlers; loggers who left heaps of slash, farmers who failed to control field burning, pastoralists who dispersed flame wherever they might prompt a green pick.

Forestry was different. While its rivals all believed fire essential to their craft, foresters denounced fire as the wood's most fundamental and implacable foe. Their ideal remained the tended woodlots and plantations of central Europe. That region, however, was an anomaly amid the Earth's climate: it underwent no routine rhythms of wetting and drying, it knew no dry lightning, it had no natural basis for fire. Although fire existed in abundance, it thrived because people put it in, and hence could take it out. (An empire forester, D.E. Hutchins, expressed astonishment, when upon visiting Germany during a study leave, he watched the smoke of agricultural burning fill the valleys of the Black Forest, a scene he thought he had abandoned when he departed East Africa.) Amid the intensively cultivated landscape of Europe, officials viewed fire as an expression of social order or disorder. Wildfire appeared during times of famine, plague, war, or revolution, when the garden went to seed and flames could feed on the waste, like rats on garbage.[7]

Its administration of reserves made forestry powerful. Through them it could claim not only the intellectual authority of science, but the political authority of the state. In country after country, foresters, as the vanguard of conservation, sought with oft-times near-fanaticism to abolish open flame. In the process they became themselves the official oracle of fire, the designated engineers of free-burning flame, the sole-source supplier of fire expertise. Landscape fire had disappeared as an intellectual subject and was not taught; incredibly enough, not even in forestry schools The dominant arrangements in Western civilisation for understanding fire — its institutions, its policies, its science — all derive from forestry's tenure as overseer of these reserves. Yet the irony is palpable: the group that most feared and detested fire became the repository of society's knowledge about fire. Even though most burning remained embedded in an agricultural context, foresters were fire's authorities, and even rural firefighting, outside of forest reserves, became their bailiwick. Within their dominion, free-burning fire was always suspect, and where possible, was legally banned and actively suppressed.

Aggressive agricultural colonisation, the countervailing demands for state-sponsored conservation, the pyric transition from a fire economy based on living biomass to one supported by fossil biomass, all these trends converged in Australia by the late 19th century as they did throughout the rest of Europe's imperium, and in Australia as elsewhere, they often focused on the firestick. Where Australia differed — where it diverged from the 'normal' narrative — was that its foresters, those on the ground, those who had grown up with a firestick-farmed land and who had to confront, year after year, the holocaust latent in every drought and desert-spawned flood of wind, refused to cast the firestick away. They determined, rather, to redirect it. As rural Australians had adapted the firestick to their purposes, so too would they. Foresters would pull it back from where it didn't belong, and point it where it did. How this occurred is one of the sagas of the Australian bush, an innovation in land management as profound as the Australian ballot in democratic politics.

Early burning, light burning, no burning

From the beginning, there were counter-voices — critics who considered forestry's fire abolitionism mad. The firefight that marked the onset of conservation thus occurred in periodicals as well as in paddocks and forest reserves. Interestingly, it typically took the form of a public debate about whether forest protection should be based on fire's control or fire's use, fundamentally the identical terms of contention that continue today. The controversies in British India and the U.S. are especially revealing, not only because they are the best documented but because they address the two political circumstances of colonial forestry. The one planted reserves amid a stubborn rural populace, while the other carved reserves largely out of a landscape stripped of permanent residents. Both confronted the scepticism of colleagues and the outright hostility of indigenes and nature.

Fire conservancy v. early burning

The story begins in India. The land was annually awash with fire. 'Every forest that would burn', recalled E.O. Shebbeare, 'was burnt almost every year'. Another forester lamented how these fires were matched 'by a most marvellous, now almost incredible, apathy and disbelief in the destructiveness of forest fires.' Under such conditions forest husbandry was impossible. Director-General Dietrich Brandis declared flatly that fire conservancy was the first duty of the Forest Department.[8]

Central-office edicts were one thing; expression in the field, another. In 1863 Brandis asked Colonel G.F. Pearson of the Central Provinces to try. 'Most Foresters and every Civil Officer in the country', Pearson observed, 'scouted the idea' and, had it failed, 'any progress in fire protection elsewhere would have been rendered immeasurably more difficult'. Pearson shrewdly selected the Bori Forest, a near-island grove of teak surrounded by cliffs on one side and a river on the other, and for the next two years his youthful corps succeeded in barring fire by laying out firebreaks, patrolling, and exhorting locals to give up burning by bribing them with goats. These were wet years, however, and fire returned savagely when the rainy seasons subsequently shortened. But foresters had seen the lesson they wanted to see, and when the Indian Forest Act was enacted shortly afterwards, 'fire conservancy' became an axiom of administration.[9]

Doubts remained, however. The first question posed at the first conference sponsored by the Forest Department was whether fire protection was possible, and if possible, was it desirable? Immediately the ranks split. Those in the field objected to fire's attempted abolition, thus joining the natives who were universally hostile to the scheme. They argued that fire suppression was impossible; that fire control only built up fuels for fires that must inevitably follow; and that fire exclusion exacted unacceptable ecological costs, smothering desired species and scrambling the fire-catalysed order of the landscape. In what might serve as a cameo, an Anglo forester (who signed himself 'An Aged Junior') described for the *Indian Forester* the puzzling situation in which, through more or less successful fire protection, the forest had acquired a tiger problem.

It is apparent that fire had not been random and ravenous, as it appeared to the British, but had been applied to particular sites at particular seasons for particular purposes by particular peoples. Those selective burns had ordered the landscape. Thanks to fire, fresh browse appeared at the proper place at the proper time; deer migrated to those sites; tigers followed the deer; and hunters knew where to find rogue tigers. But eliminating fire, or smearing it, affected that land as the abolition of caste would Indian society. Boundaries blurred. The ecological order became confused. Tigers no longer kept to their place — their place being scrambled and overgrown. They began to menace local communities, follow rangers, and generally make themselves 'disagreeable.' The forest now had 'much fire conservancy and many tigers'. Whether successful or not, the *attempt* at fire control was sufficient to unbalance the Indian biota. Changing from small fires set annually to large fires that came every three or four years did not preserve the old order. It was not simply fire that India needed, but its syncretic order of fire regimes.[10]

In practice, only the most valued blocks were spared fire, and this at the cost of protective burning all around them. In some locales, foresters discovered to their chagrin that native guards were the largest source of fire, this because their pre-emptive protective burns were the only means to shield the core forests. Worse, where more or less successful, fire exclusion retarded regeneration in teak, sal, pine, and bamboo, precisely the species most valued commercially. The natives' burning practices had yielded the woods Britain wanted; its own practices choked out those trees in favour of combustible 'jungles'. The critics wanted a regular regimen of burning, beginning with early-season fires. Rather than 'fire conservancy' based on fire suppression, they proposed 'early burning.'

Against them stood the authorities: the urban-based administrators, the academics, the whole apparatus of officialdom and technical expertise. Imperial rulers did not declare a cause and then withdraw; it just wasn't done. Loss of control over fire meant a loss of control not only over the woods but over the natives who set those burns. Global forestry — a bold, dazzling vision of conservation — had defined itself by promising to tame both axe and fire. The reach of that vision was astonishing. (Rudyard Kipling even wrote a sequel to the *Jungle Book* in which he has Mowgli, having grown up in the man-village, work as a forest guard for the Forest Department, giving 'sure warning' of and fighting jungle fires.) The whole ideology of political rule, that the new order represented reason and modernity against superstition and rutted tradition, would turn to dust if those premises were false.[11]

But the situation on the ground got worse. Wildfires became fiercer, locals resorted to arson, forest guards surreptitiously burned, and the prime timber refused to regenerate. Critics argued for a hybrid program in which controlled burning supplemented fire suppression. In 1897, inspector-general Berthold Ribbentrop had to intervene. To protect young trees and forest litter (the twin obsessions of European forestry) — to say nothing of saving imperial face — he ruled for the further expansion of fire conservancy. Edicts did not suppress doubts, however, or sub-rosa burning, and by 1902 the debate flared across the pages of the *Indian Forester* and the annual reports of conservators. In 1905 a compromise was proposed by which controlled burning could be brought into working plans.

The climax came in 1907 when the Conservator of Forests for Burma simply withdrew. The natives, with their prodigious burning, had grown teak; the British, with their hard-fought commitment to fire conservancy, had not. By 1926 'early burning' had acquired the imprimatur of the *Indian Forest Manual*. While imperial foresters, and other officials, never liked the practice, they recognised that if they were going to cultivate native species with the aid of native guards, they would have to adapt native burning as well. The best way to counter indigenous fires was to impose their own. As a chronicler of the controversy, E.O. Shebbeare, concluded that, for all the various ills of fire conservancy, 'fire appears to be the only real cure'.[12]

Systematic fire protection v. light burning

In America the debate began almost as early. John Wesley Powell had written vigorously about fire in his 1878 *Report on the Lands of the Arid Region of the United States*, along with George Perkins Marsh's *Man and Nature* one of the seminal texts of American conservation. The fires threatened the timbered catchments on which irrigation depended; they were set by the Indians; controlling those indigenous fires would promote a rational settlement. A careful and sympathetic observer, Powell became head of the Bureau of American Ethnology and later a celebrated chief of the U.S. Geological Survey. By 1890, at the height of his political influence, he reversed himself and argued for an adaptation of indigenous burning on the grounds that it helped prevent damaging as opposed to extensive fires. Likewise he argued for locally controlled, watershed-based conservation entities rather than the imperial model of forestry. Foresters, of course, were outraged, denounced his scheme as vandalism, and dismissed his burning proposals as mere 'Paiute forestry,' the Paiutes being a desert tribe Powell had particularly studied but who were technologically primitive, living on seeds, grasshoppers, and rabbits.[13]

Among those on the ground, however, enthusiasm remained for surface burning. A famous account for a California forest reserve survey in 1904 observed that 'the people of the region regard forest fires with careless indifference'. Even to 'shrewd men', the fires

> seem to do little damage. The Indians were accustomed to burning the forest over long before the white man came, the object being to improve the hunting by keeping down the undergrowth, which would otherwise shelter the game. The white man has come to think that fire is a part of the forest, and a

> beneficial part at that. All classes share in this view, and all set fires, sheepmen and cattlemen on the open range, miners, lumbermen, ranchmen, sportsmen, and campers. Only when other property is likely to be endangered does the resident of or the visitor to the mountains become careful about fires, and seldom even then.

America then was much like Brazil today. Foresters found the festering fire scene an endless outrage.[14]

The issue came to a head during the Big Blowup of August, 1910. In newspapers and periodicals, a strong chorus of critics, mostly in the west, denounced the fire-suppression policy of the Forest Service, which they insisted would only lead to worse fires, diseased woods, and greater expense. One critic even suggested the Army be called out, not to fight fires, but to light them. Instead, they proposed the frontier's favoured alternative: 'light burning,' or 'the Indian way' of forest management. The controversy split even the two major departments of the federal government that oversaw the public lands. The Secretary of the Interior backed light burning; the Secretary of Agriculture, who had jurisdiction over the Forest Service, sided with fire control.[15]

The fight continued until the mid-1920s when a board of foresters officially condemned light burning as heretical and dismissed its proponents as the conservation equivalent of perpetual-motion mechanics and spoon-bending psychics. Government foresters revived the slur of Paiute forestry, believing that light burners — who did not have 'professional' standing equivalent to academic forestry — were merely pawns or fronts for timber companies, ranchers, the railroads, and others openly hostile to forestry and secretly intent on plundering the public domain. In fact, timber-owners often practised light burning, as did the Southern Pacific Railroad on its forested lands. No less suspiciously, the best known public voices for light burning were literary figures such as the poet Joaquin Miller and the novelist Stewart Edward White, who were scorned by academic forestry as self-evidently incompetent. The battle was bitter; but foresters had the public lands and the force of government behind them and prevailed.[16]

In retrospect, the arguments seem perverse. Foresters wanted exactly the thick regeneration and scrub that a later generation would identify as stoking uncontrollable fires. Two powerhouse personalities, S.B. Show and E.I. Kotok, described the situation that had evolved in California. When the national forests were established, everyone engaged in forest burning, which was an 'established practice.' The idea that 'fires could be excluded entirely from millions of acres was generally regarded as preposterous'; critics prophesised all kinds of ills were it to be attempted, not least 'the uncontrollable crown fire.' As fire protection strengthened, the amount of flammable material, as forecast, 'increased greatly'. That prompted locals to take to burning — some as incendiarists, some as light burners, but both were equally threats 'not only because of their direct action, but even more so because of their open preaching of fire'. Forty years later the authorities were the ones preaching fire restoration and warning about unmanageable fuels.[17]

A similar story characterised the country overall. In the

southwest Aldo Leopold, the towering figure of American environmental thinking in the twentieth century, then a graduate of the Yale School of Forestry and busy establishing a fire-protection program for the Forest Service, condemned light burning and warned that it directly threatened the policy 'of absolutely preventing forest fires insofar as humanly possible'. In the South an exasperated Forest Service turned to fire-prevention campaigns modelled on evangelical revivalists ('the Dixie Crusaders'), convinced the American Association for the Advancement of Science to organise anthropologists to study woodsburners as though they were natives from previously undiscovered islands, hired a psychologist to inquire into the locals' superstitious resistance to forest science (so unreasonable were their actions and laughable their explanations), and finally suppressed field evidence that contradicted dogma, not only because it was toxic politically but because they knew, just knew, the data had to be wrong. Everywhere, European forestry sought to suppress indigenous fire practices, and everywhere the issue was fraught with cultural and political connotations.[18]

Still, the issue wouldn't go away. By the 1930s other professionals in regions of the country not dominated by federal lands renewed the protest. Wildlife biologists found that prime hunting species were vanishing along with their fire-maintained habitats. Agronomists recognised the value of fire-freshened pasture. Commercial foresters appreciated the necessity for regular underburning as a means to clean out the 'rough' and contain wildfires, and eventually select academic foresters understood that without proper burning some critical species, like the longleaf pine, would not regenerate. These issues became public collectively in 1935, just as the U.S. Forest Service was preparing to announce its final assault on free-burning fire in the form of the 10.00 a.m. policy, which stipulated a single, national standard for the control over every fire (by 10.00 a.m. the morning following its sighting). Once again, the larger politics and an unholy dose of professional stubbornness pushed suppression.

By the 1960s, however, the weight of evidence in favour of fire both for fire control and ecological health had become overwhelming. Beginning in 1962 the Tall Timbers Research Station, a private facility, gave critics a forum outside government control (they refused to submit their proceedings to peer review on the grounds that censorship on fire was pervasive). In 1963 the Leopold Report for the National Park Service used fire exclusion as the poster child for mismanaged parklands. And in 1964 the Wilderness Act provided legislative leverage to reintroduce a natural process to natural areas.

Quickly, the old hegemony crumbled. Policy was reformed to allow, and furthermore to encourage, fire's reintroduction. It is important to recognise that the critical developments took place in the southeastern U.S., for which the Forest Service had limited holdings and in which a tradition of rural burning had not been wholly extinguished; that professions other than forestry and bureaucracies other than state-sponsored forestry made the case for burning; and that the general public became interested not through fire itself but through the cultural magnetism of wilderness. A reformation in fire policy was part of a larger social revolution in environmental values; within a decade, for example, the

Endangered Species Act was legislated. Collectively, such acts became wedges that split open the administration of public lands.

All this, fire assimilated. As people changed land use, so fires changed as well, and as fires made their own demands, so administration adapted, or if not, they had to confront a land out of synch with burning. The fires they wanted they couldn't get, and the fires they feared came without regard for the awkwardness of agency reorganisation and the bureaucratic bromides of good intentions.

And Australia?

India's early-burning debate was virtually indistinguishable from America's light-burning controversy, and both strongly resembled Australia's chronic muttering about burning off. One Australian forester confessed that light-burning was so 'practically similar' to what he confronted that he dismissed the eccentricities of the Australian scene as 'not vital to the issues involved, nor do they affect the conclusions arrived at'. What distinguished Australia was not its choice one way or another but its absence of choice. It looked to both the Indian strategy and the American strategy without accepting either or, for a time, concocting a third way of its own.[19]

As a member of the empire, Australia belonged on the imperial circuit of state-sponsored forestry. Its administrators trained in the same curriculum and at the same schools as did those elsewhere, and were subject to the same doctrines. In

1914 a grandee of that school, Sir D.E. Hutchins, following an extensive tour of Australia, wrote at length about Australia's vaunted bushfire problem, which he scotched as the product of slack organisation and weak will. The forester, he declaimed, was 'a soldier of the State, and something more'. If the south of France could control fire, so could South Australia; if Cape Colony and India could come to grips with indigenous burning, so could Victoria and Queensland.

So, likewise, Australian foresters followed closely the American scene. The muddled outcome was illustrated in 1923, when Owen Jones informed the 1923 Imperial Forestry Conference that Australians 'use fire to prevent fire'. He had no doubt that 'it seems a very wrong and a very dangerous thing to do', but under Australian conditions it was both useful and necessary. The Conservator of Forests for Western Australia soon afterwards reprinted the 1923 condemnation of light-burning by the California forestry board.[20]

A year later he summarised the issue as it related to Australia. 'Two schools of thought' have emerged, Stephen Kessell argued. One believed that the only solution to fire control was to clean up the forest floor and set frequent light fires to suppress fuels; the other, that 'absolute control of all fires in and around the forests is the first and biggest step towards the establishment of Forestry'. It was the 'practical forester, of the woodsman type,' another forester recalled, who 'preached the need for controlled-burning.' To professional foresters, particularly those trained in Europe, 'this was heresy'; they insisted that fires would vanish as better forestry came into practice. State foresters might have to wink at a bit of controlled burning while they wrestled the lands and society into proper shape; but once established, those lands would themselves help ward off flame. The ultimate solution was to convert the 'present wilderness' into 'tended forests'.[21]

Until then, with burning continuing as an expedient, Kessell took the logical step and sought to rationalise it. In the prime-timber jarrah belt in south-western Australia, he demonstrated to his satisfaction that fire control was 'economically possible under conditions obtaining in this State.' That model he sought to project throughout the forests under his dominion. 'Complete fire prevention is impossible in practice', he concluded, 'and controlled fires play an important part in silvicultural operations'. They were useful for clearing away slash left by logging and for stimulating regeneration. Still, burning was not 'a cheap and easy method of solving or dodging the fire problem'. It was unclear how far beyond those felled coupes and high-timber sites it might be pushed, or whether the techniques, as an authorised instrument of state forestry, might come to light elsewhere in Australia, much less the world.[22]

In the end, Kessell's burning experiments remained a niche practice. They didn't challenge prevailing paradigms or transcend their pedigree in the jarrah. The utopian vision inherited from central Europe of a fire-free forest endured, like a mirage ever looming in the horizon. The value of controlled burning as a working, temporising technique was unquestioned in the field. On the grand controversy over whether to light fires or fight them, Australia found itself in both camps, and in neither. Its forestry lacked the intellectual cachet of Europe, and the economic clout of Canada and America. Its foresters observed, experimented, and muddled through. Besides, Australia was rife with peculiarities, though none so strange that they undermined the grand principles. If its trees shed their bark, they could nonetheless be milled, and so, too, if its foresters did a bit of burning, they knew the proper purposes. If the time had not yet come to impose an appropriate order on the antipodal continent, that day was approaching.

Between two fires: creating an Australian strategy

Until 1939 Australia followed these prescriptions. In practice, Australian foresters, like their brethren elsewhere in the empire, accepted compromises with fire, while insisting on the purity of their principles, that land was better without burning. The Black Friday holocaust, paradoxically, forced a change. The royal commission headed by judge Leonard Stretton listened to locals as well as to officials, and found cause to condemn both. The conflagrations, in Stretton's celebrated phrase, 'were lit by the hand of man.' Yet he recognised the necessity of fire for those on the land, and the absurdity of forestry's belief that the land would ultimately abolish fire if fire could be kept out for a sufficiently long time and the quixotic assertion by foresters that they could accomplish that prohibition. The problem, Stretton concluded, was not controlled burning, but badly done — that is, uncontrolled — burning.

Big fires become great fires when they strike against other cultural concerns. A colossal one was about to hit Australia: World War II. H.G. Wells improbably found himself in Canberra when the 1939 fires struck, along with other savants from the Australian and New Zealand Association for the Advancement of Science.

'The younger men', he noted, 'went off firefighting.' The elders toured the scene, and Wells used the conflagration, presciently, as a metaphor for the coming war and the defences Australia would require as its prime population went to fighting. The war folded fire defence into national defence; other fires struck in 1942 and 1944. Stretton led a second royal commission for the 1944 crisis, and followed with another in 1946 that looked at mountain grazing. Fire — used well or ill, exploited to enhance or degrade — formed a focus for his great trilogy of commissions. Between them, the war and Stretton's commissions ensured that the 'hand of man' and the fires it kindled would not fade from public consciousness as others had.[23]

The post-war era brought a new generation of foresters, more openly nationalistic, more sympathetic with rural mores. Another outbreak of fire in the alpine region during the long summer of 1951–52, sites now absorbed into the Snowy Mountains Scheme, galvanised them into reform. Australian forestry then did something that no other industrialised country did: it adopted controlled burning, not fire suppression, as the basis for the protection of its wildlands. The tributaries to this decision were many: recognition of the manifold adaptations of the indigenous biota to fire; appreciation for the long traditions of burning by Aboriginal peoples; acceptance that rural burning had a value; admittance that Australia could never afford a paramilitary campaign against fire like that emerging in North America; and a determination that Australia could craft its own, unique strategy, neither a European echo nor an American analogue. From Western Australia came the first system for integrating controlled burning into public forestry; from New South Wales came the idea of cooperative fire protection by multiplying and coordinating volunteer bushfire brigades. Forestry brought the intellectual rigour of science to rural beliefs, and the political discipline of state institutions to rural practices. Then, in 1961 bushfires blasted Western Australia, incinerating the town of Dwellingup. Under G.J. Rodger, a former director-general of the Commonwealth Forestry and Timber Bureau, a royal commission investigated the disaster and gave the revolutionaries an official imprimatur. The future lay with systematic, expansive, hazard-reduction burning.

Between these two fires and the royal commissions they inspired — the 1939 Stretton Commission and the 1961 Rodger Commission — emerged an Australian strategy that its nationalist foresters unabashedly presented to the world in defiant counterpoise to the 'North American' model. An official announcement quickly followed in 1962 when Alan McArthur published *Control Burning in Eucalypt Forests* through the Commonwealth Forestry and Timber Bureau. By the mid-1960s Australia's fire community was proudly devising aerial incendiaries to ensure broad-area burning: the small Cessnas dropping tiny capsules of flame contrasted nicely with lumbering American air tankers drenching fires with chemical retardants. The one stood for a pragmatic, slightly sardonic understanding that controlled burning, both inexpensive and indigenous, had to remain the essence of Australia's strategy of fire protection; the other, for an expensive and ultimately futile juggernaut fuelled by loose money, war-surplus hardware, and a Cold War mindset toward the 'red menace.' Unapologetically, Australian foresters saw themselves as nimble, clever, bold advocates for a fire realism distinct from the conventional, if heavily endowed, wisdom that they regarded as wrong for Australia and probably misguided most everywhere else.

The firestick had passed hands. It had gone from lightning to Aborigine to rural settler, and now to public forester. Call the outcome 'firestick forestry'. In each transfer controlled fire had interacted with new livelihoods, and thus yielded different outcomes: there was no single way that fire might occur on the land; what survived was the imperative to find the right kind of fire for the required habitat. Even the most ardent advocate for the Australian strategy never insisted that every hectare be burned to dampen fuel, or that hazard-reduction burning could serve, by itself, as a complete system of fire protection. But systematic burning, informed by science and restrained by bureaucracy, could lessen damages, improve chances for fire control, reconcile Australians with their environment, and not least project a distinctive Australian identity to the world. That was no mean feat, and the architects of the Australian strategy knew well the breadth of their achievement.

This was their heroic age. The earthly epic that was fire could now speak with an Australian accent. Between the two major fires — Black Friday and Dwellingup — the founders had a creation story, a founding saga with a stature somewhere between the *Song of Roland* and 'The Man From Snowy River'. In the patriarchal A.G. McArthur they had a Moses, with a firestick for a staff, resolutely guiding a fickle people to a promised land visible from an antipodal Pisgah that seemed to reside somewhere in the Snowy Mountains Scheme.

Importantly, those two framing fires spoke with the authority of royal commissions, themselves mythic in ambition and memory. Judge Leonard Stretton was their Homer; G.J. Rodger, their Virgil. What they did not appreciate, amid the euphoria of revolution, was that creation stories typically end as tragedies.

Backfire: the environmentalist critique

It didn't turn out the way they planned. The reasons are many. Australia became even more urban, or suburban, or exurban, shunning its rural heritage as a foundation of national commerce. State-sponsored forestry, which had thrived as a global enterprise, went into a global recession, and eventual collapse; its estate parcelled out to new purposes and its commodity economy condemned. An institution that had long prided itself as the vanguard of conservation now found itself denounced by environmentalism. The industrial transition intensified, driving open flame ever further from the daily lives and consciousness of Australians, save when standing over their barbies or when feral bushfires burst upon them. The reformation wrought by the post-war generation turned out to be a transitional moment, not a final state of equilibrium. With eerie irony, foresters now occupied the position that graziers had in 1939. Not surprisingly, the critiques turned on the most vivid and dramatic of their practices: deliberate burning.

Hazard-reduction burning — firestick forestry, the core of the Australian strategy — seemed to New Australia less like the herald of a bold new nationalism than a lingering anachronism from a feckless past they were eager to shed. High-intensity regeneration burns on logging slash looked like mechanised selector burns, while broadcast burning appeared as little more than industrial strength burning-off, a rural relic no different from mindless land-clearing, irrigating, and poisoning, a kind of vandalism wholly out of synch with notions of wilderness, biodiversity, and ecological integrity. The aerial ignition that foresters celebrated was denounced by environmentalists as wanton, a dark symbol linked not to the Aboriginal firestick but to American firebombing in Vietnam. Much as foresters had sought to restrict farmers and graziers by shackling their firesticks, so an emergent environmentalism sought to shut down state-sponsored forestry by denying it the essential tool of its enterprise.

Environmentalism proposed other uses for Crown land: more recreational sites, more nature preserves, some Aboriginal places, and less logging, wood-chipping, and grazing. Critics wanted public lands moved out of forests and into parks; they wanted state-sponsored ecology to replace state-sponsored forestry; and so, too, they wanted the firestick transferred to hands they trusted — the hands of people more like themselves. Control over fires mattered less than control over a fire-fighting bureaucracy whose ultimate goals they considered suspect. Broad-area burning seemed less likely to preserve valued species and habitat than to homogenise them: it would destroy what agencies publicly declared they sought to preserve. Fire protection meant protection of biological values from fire agencies and their ill-advised burning, not just the protection of economic assets from wildfire.

State-sponsored forestry, which had spiralled upward with

global colonisation, now spiralled downward with global decolonisation. Unsurprisingly, everything they did was condemned, not least their fire programs. In most of the world, eager to promote the restoration of natural conditions, including fire regimes, this meant condemnation of state forestry's fire-suppression juggernauts in favour of fire-restoration schemes. In Australia, the critique got turned on its head, and its foresters, who had broken away from the global mob, found their distinctive creation scotched along with the rest. Like mountain graziers before them, foresters argued that fire was necessary, and that without regular burning ever more savage bushfires would erupt. And like those forlorn graziers, they found themselves largely denounced as ravagers of the environment, ignorant of contemporary values and indifferent to the wishes of modern Australia. They hoped their good fire practices would help justify the other things they did; instead, the logging, grazing, and roading only tarred the burning. Politics and professional competition were not far below the surface. Foresters insisted that, because they knew fire, they knew how to manage the land overall. Greenies insisted, with equal conviction, that because they knew how to manage the land better, they knew fire as well.

Controversy kept the issues simmering and conflagrations put it to boil from time to time. Gradually, two trends emerged, one divesting state forestry of its authority over rural fire, and the other, its jurisdiction over its own lands. Collectively, they challenged forestry's strategy for managing fire, the domain in which it might exercise that strategy, and the on-the-ground power it could apply to that scheme. State-sponsored forestry found it had less reach over the countryside and a lessened grip over what it still grasped.

The first trend began to transfer responsibility for crisis fires to an emergency services bureaucracy that could coordinate the various fire-fighting resources around a state — volunteer bush-fire brigades, the machinery and labour available through industries such as logging, the fire-control capabilities on state forests. While foresters remained a vital component, their authority for general rural fire protection withered. And as land use shifted away from industry, the resources that logging, in particular, had made accessible for fire suppression also left the scene. Rural fire, too, changed character: it became more exurban. Fields and paddocks were less at risk than the fringe of a fractally expanding urban bush. It was not clear that such residents, often commuters or seasonal residents, could staff the volunteer brigades that furnished the backbone of rural fire protection, much less that they would be enthusiastic about regular burning and its annoying smoke. Eventually the capacity to fight fire might remain constant, or even expand, but the transitional era was one in which the old apparatus was removed before a new one could be installed.

The second trend struck even more deeply. State forests themselves were disestablished by gazetting them as parks or preserves or by having the commercial chunks subject to new regulations or reallocated to a public corporation. Again, the effect was to shrink the dominion of state forestry. The peculiar accomplishment of the heroic age — that it had gathered into a common realm of fire management the varied concerns of Australia's bush, both rural and public — disintegrated. State foresters found their capacity to fight fire crippled, and their understanding of how to manage fire, challenged. All these issues, and others, swirled around the emblematic core of the Australian strategy, hazard-reduction burning. The heroic generation had devised a program suited to a rural Australia that no longer existed. Critics saw the project much as they might look back on the Snowy Mountains Scheme, as a misguided nationalist boast that had encouraged poor land-use in an arid country. So, too, they viewed Australia's misplaced fire mismanagement as threatening biodiversity as much as its misplaced water mismanagement had soils.

The country was changing. A New Australia was eager to turn even the fabled Red Centre green; this time not with irrigation but with more biocentric values. State-sponsored forestry found itself on the wrong side of that fence, paradoxically allied with its erstwhile rivals, its fire strategy considered of a piece with salinated wheatfields, eroded high-country pastures, and wood-chipped native forests. Instead of accruing honour as inventive and courageous nationalists, they were vilified as isolating Australia from the axioms of global greenery. Stunned, foresters have protested, and predicted the return of wild fire if fire control did not bulk up its capacity to strike small fires and if their prescriptions for broad-area burning were ignored.

PART III

As the world burns

Over the past decade the forecast fires have come, and come again. Each outburst has sparked inquiries: the literature of those reviews constitutes almost a genre in itself. For the Alpine fires of 2003, there were some seven official inquiries (and counting), plus innumerable submissions, critiques, and commentaries from professionals and laity alike. Yet until that outbreak, despite the media frenzy surrounding the bushfires, the larger narrative, however counterintuitive, remains a chronicle in which fire is receding on the Australian landscape. The area burned by bushfire continues to shrink, as does the land burned under prescription. This backdrop is what made the 2003 season, which nearly replicated the scale of 1939, so shocking.

This fiery resurgence was not unique to Australia, however. Nearly every place endured what has seemed an unprecedented plague of conflagrations: from Mediterranean Europe to the Russian Far East, from Kalimantan to Brazil. Since the 1988 fires, when alarm over burning in Amazonia coincided with a tsunami of flame that washed over America's Yellowstone National Park, the media have become sensitised to backcountry burns; and where the fires overrun famous places, fire reportage takes on the character of celebrity reporting.

The U.S. suffered through a tormented decade of fire, from

1993 to 2003, that commenced and concluded, full circle, in Southern California. A tenacious drought brought waves of fire in the American West in 1994, 1996, 2000, and 2002, while Florida burned fiercely, with over 100,000 people evacuated, in 1998. The 2000 fires in the Northern Rockies were the worst in 50 years, and harked back to the epic Big Blowup of 1910. By 2002 the cycle had spawned the biggest fires of the century in the states clustered in the Southwest. Conveniently, the fires coincided with national election years, and became part of the lead-in to the autumnal vote. Burned-over crews, ravaged suburbs, concern over fire ecology — these informed dramatic proclamations and sparked new reforms. What happened in Australia was thus not unique. Its blazes merged into what has seemed a global rampage of burning.

The fires appeared like a fiery version of the emergent plagues that catastrophists were forecasting at the time, and like them, their causes reflect an interaction of people and nature. Trying to tease those two generic causes apart, demanding that one or the other, either nature or culture, must dominate is a metaphysical pursuit, like insisting that physicists decide once and for all whether an electron is a particle or a wave. It is neither, and both, depending on how one frames the experiment. So it is, also, with wildfires: it is impossible to disaggregate nature and humanity from the Earth's contemporary fire scene.

Contributing causes have assumed a kind of hierarchy. The immediate prods are the usual suspects — drought, wind, fire-favouring terrain, a well-timed spark. Yet the deep drivers remained unaltered. This is a world whose gears and pulleys connect to the machinery of industrial combustion, a force so powerful and subtle that it can destabilise the global climate. To the extent that violent swings of drought, in particular, come from a forced warming due to fossil-fuel burning, this condition can even pull the levers of the Rube Goldberg machinery that constitutes the Earth's weather and that, for a period of days or weeks, sustain the eruptive megafires. Similarly, industrial fire, broadly conceived, is underwriting two drivers of more intermediate status. One is an increase in the stuff available to burn, and the other, a change in the capacity of people to assist or retard that burning. The first reflects modifications in land use; the second, of institutions established to oversee fire's presence.

Stoking the fires: more fuel

There is more stuff to burn. Drought accounts for part of the new load. Dry material burns better than wet, and a deep drought makes large chunks of a countryside available for combustion that normally could not kindle. It matters not whether the 'fuel' is native flora or logging debris, a stubborn dry spell can transform it from heat sink to heat source. A decade of intense, migrating drought has liberated large patches of normally non-flammable or episodically flammable forest for burning — a decade that echoed the Long Drought a century before. Had those years been wet, the fires could not have surged with such savagery. That was nature's inestimable contribution.

But people helped shaped the rude biomass on which climate could act. The most dramatic build-ups followed migrations, of which two variants are most prominent, one peculiar to developing countries and one to developed. Agricultural colonisation in Borneo and Brazil, for example, has slashed normally fire-immune forests into kindling; widespread fires in eastern Kalimantan and eastern Amazonia followed upon frontier land-clearing and logging, an historical echo of what had occurred in North America and Australasia a century earlier. A rural exodus in Greece and Portugal into metropoli such as Thessaloniki and Oporto allowed the residual vegetation to erupt and overrun the countryside. In such fire-flushed lands, only close cultivation, imposed over thousands of years, had traditionally checked flames, but now those fussing hands had gone to the cities, the snipping teeth of goats and sheep were fewer, those who remained to handle the fires were elderly, and the countryside fluffed into tinder and flame.

The inverse process, found in more industrialised nations such as North America and Australia, was an exurban recolonisation of rural lands. Here once-cleared landscapes returned to scrub, while newcomers, often retirees or recreationists, stuffed the landscape with wooden structures that appeared to fire as no different from logging slash.

This frontier of settlement, like its predecessor, has encouraged wildfire; not, as previously, because it chewed up the countryside into kindling but because it left the bush to blossom unchecked. An odd symmetry emerged between First World and Third. In the latter, those at risk were the poor and powerless, pushed to the dangerous margins; in the former, they were the rich and mighty, choosing the fringe as the more desirable of estates. It was, after a fashion, the realty equivalent of extreme sports and a reflection of similar social values. Common to all, however, was rapid movement — the abrasion of people against land, the strike of urban steel against bush flint — that led to ignition. Flame feasted on change.

The truly malicious irony, however, was reserved for the most fire-protected bush, the public domain. Here decades of attempted fire exclusion in the name of good government, right thinking, and the protection of timber and catchments had let nature stockpile combustibles like a bower bird seeking to entice flame. What had, in the past, been trimmed by axe, tooth, and firestick now overgrew sites, and with the kind of combustibles that can feed fires of unprecedented ferocity, often well beyond the range of evolutionary experience of either the indigenous or exotic flora. The effort to abolish fire had not extinguished flame but only forced it to metamorphose into other forms. Like levees thrown up to protect a floodplain, control efforts had eliminated small, nuisance events, only to worsen the setting for the massive ones. Instead of routine fires washing away the encrusting fine fuels, catastrophic fires overflowed and drowned whole forests.

In such ways, climate and combustibles, with fiery synergy, sparked troubling burns.

Agency and accountability

The other half of the equation, however, was a change in how people responded to those eruptions. In some cases, political disorder broke the capacity of nations to prevent or suppress wildfire. This was the case, for example, with Russia and especially Mongolia, where social disintegration following the collapse of the communist state led to a rash of fires while bureaucratic breakdowns stalled the ability of society to contain them. Fire-fighting forces were dissolved, selectively withdrawn, or crippled. With less geopolitical drama, this was also the case in countries like Portugal, where internal migrations outpaced institutional responses, leaving a lethal vacuum in fire-primed landscapes no longer subject to customary practices but not yet absorbed into an alternative arrangement. Fires that might have been caught early, or never started at all (much less recklessly or maliciously), blew up into conflagrations.

Elsewhere, the shift has been more subtle. It could, for example, take the form of a policy change that sought to limit suppression and encourage burning. It might mean a transfer of lands from commercial forests, subject to intensive protection, to nature preserves, with a kinder, gentler standard of control. Or it might appear as a division of duties, with fire protection diverted into a more generic emergency-services bureau. Overall, in the name of greater economies, there might well be a downsizing of protection forces, a diminution of fire staff, a divorce between fire services and land management. Such administrative reforms affected how institutions dealt with fire, especially those with oversight for the public estate.

Compared with the planetary tides of the Southern Oscillation or the continental force of south-east Australia's 'fire flume' (as one author has termed the geographic funnelling of desert winds and flame) such matters might seem trivial, but that perception only demonstrates how feebly we recognise ourselves as fire creatures — indeed, our reluctance to admit to our power (and responsibility) as fire monopolists. There is no 'fire problem' in the world that would not vanish if people willed it, either by changing behaviour or proclaiming different standards. People alone cannot of course make fire come and go at will, save in select environments, but there is no environment, even the most fire-prone and naturally fire-informed, that people cannot refashion.

The perception is complicated as well because the changes often seem tiny. But a new policy or a new idea can set off huge outcomes, the way a miniature solenoid switch can turn off a massive engine. Most of the institutional reforms in America and Australia began in just this way. In some cases, there was a belief that fires did less harm than fire-fighting. In others, a conviction grew that let-burning could bond ecology with economy such that fire management could be both less intrusive and much cheaper. If fire was to be tolerated, or even encouraged, so the reasoning went, there was little need for the opulent apparatus of mechanised fire-fighting and stand-by staff. The exclusion of woods workers, too, removed both crews and equipment that fire organisations could once tap during emergencies. And, not infrequently, there was simply confusion. It was unclear how to translate policy into practice, and as with any period of transition, the agencies were vulnerable. Often it was unclear what to do, who should do it, or how it might be done.

Unlike the scene in the former U.S.S.R. and its client states, the forestry bureaus that housed fire protection were not shattered but split, transferred, realigned, renamed, rechartered, and reorganised — often over and over. Everywhere, state-sponsored forestry found itself devolved, partially privatised, disaggregated, or even, as in New Zealand, disestablished altogether. In the United States the national Forest Service was effectively gutted, like a heritage building, and retrofitted with new innards; new policies, personnel, purposes. While its name and exterior survived, the U.S. Forest Service no longer functioned as before, and was keen to proclaim that new identity to a sceptical public. Forest 'health' and reinstated fire, not timber and fire suppression, were its cardinal missions. We will restore fire where we can and fight it where we must, the chief forester assured the public. What one might call the imperial model of fire conservancy, the legacy of Europe's colonial conservation, died on its feet.

These institutions, however, had been the bearers of fire protection, not only on public lands but typically in the rural landscape as well. In most parts of the world, having wrested the firestick away from the general public — as though open flame needed to be a government monopoly, like atomic energy — these agencies had become indispensable for general bushfire protection. Even where rural volunteer brigades existed, state forestry bureaus created an interstitial medium by which the separate institutional parts could connect. This was true even in Australia, rightly celebrated for its extraordinary network of bushfire brigades. Anything that ruptured those agencies could ripple widely through the periphery.

PART III

The new world order on fire

Climate, fuel, and institutional uncertainty — an ideal formula for a decade's perfect storm of fires. What happened in Australia has clear cognates elsewhere. Fires in Australia's urban bush mirror those in the 'wildland/urban interface' of the American West and the spreading recreational fringe of the Mediterranean littoral. Megafires marauding through landscapes once thought reasonably protected have become a new norm: in Montana and Arizona, in Alberta and British Columbia, no less than in the Blue Mountains and the Otways. The transfer of lands from state forestry to nature reserves, accompanied by shifts in policy, has caused fire agencies to stumble for the past three decades. Yellowstone National Park in the U.S., Kruger National Park in South Africa, Alpine National Park in Australia — all flagship institutions, all overrun with fires that swept from 45–75 per cent of their holdings in one wild surge.

The drive to reinstate a proper regimen of fire has everywhere proved costly, embarrassing, and even lethal. A 200-hectare controlled burn in Wilson's Promontory blew up and scorched 6,000 hectares; a 200-acre prescribed burn at Bandelier National Monument in the U.S. exploded over 45,000 acres and scoured the town of Los Alamos while a sister burn at Grand Canyon forced the evacuation of the North Rim and was halted only at the Canyon's brink. Droughts, excess fuels, unstable agencies, big fires — these were not unique to Australia.

What differentiated Australia was that it, alone among

the developed nations, and particularly among the Big Four firepowers, had institutionalised controlled burning within its public estate. To many Australians, particularly its metropolitan populace and its educated elites, that often seemed an embarrassment, if not an outrage, like continued grazing in the mountains or poisoning for pests. Among its firepower peers, however, that achievement granted Australia considerable stature. The others saw in the survival of the forester's firestick an opportunity that they, to their grief, had lost and sorely sought to regain.

Fire's rectangle: options for management

Traditionally, the fire community casts its ideas into groups of three, for which the venerable 'fire triangle' — the combustion trilogy of heat, fuel, and oxygen — is the archetype. But a century of experience with fire protection on the public estate has happened in enough countries to abstract a few principles of administration and strategy, and these seem to group naturally into four. (This may not be as radical an innovation as one would like to believe. There is an old joke about the 'fire rectangle' which plays on the training adage attached to the fire triangle, that the ambition of fire control is to break the linkage among the parts. Remove any side of the triangle and the fire goes out. The rectangle adds a fourth element: supervisors. Remove any side and the fire goes out.)

This acquired wisdom applies to fire-prone public lands. There are public lands that are not naturally susceptible to burning; they undergo no cycle of wetting and drying, have no experience with dry lightning, are not salted with pyrophytic plants. And there are plenty of privately held lands that can burn with a vengeance. On such sites, however, the range of options is greater, notably, that there is no political requirement to maintain the land in a quasi-natural state; rather, the incentives are to remake the site to suit the economic or aesthetic interests of the owner. One could eliminate fire from the scene by replacing bush with golf courses, suburbs, and shopping malls. If Bill Gates bought a chunk of the Blue Mountains or Disney Corporation the Dandenongs, one could expect a very different regimen of fire. But public forests, parks, and monuments don't have those options. What they can do is let fire burn, suppress it, substitute tame fire for feral fire, or change the combustion characteristics of the landscape so that fires of any sort burn in ways people prefer.

Let-burning

What was once a reluctant necessity has become today an avid virtue. If the land is to be left in a natural state, then its care should be left to nature.

In the early days, fire agencies let fires ramble in the backcountry because they didn't have the wherewithal to fight them. As the agencies 'developed' the land with roads, trails, lookouts, and the like, they attacked those fires under the hard-learned doctrine that a fire hit early was a fire more easily and cheaply contained. Fires allowed to smoulder smoked in valleys and hid new fire starts; and some fraction of those loitering fires eventually went looking for trouble. The longer a burn lingered, the more likely it was to encounter a stubborn dry spell or a stiff wind and blow up. Since it was unclear which fires might go bad, the surest strategy was to hammer them all. The way to prevent big fires was to extinguish all small fires. The cost might seem exorbitant, but only until one innocuous burn metamorphosed into a monster that could consume budgets as completely as it did the bush.

Bushfire protection thus resembled urban fire protection: rapid detection and initial attack were the essence of sound fire control. As public land agencies matured, they pushed their fire protectorates further and further into the backcountry, a kind of forward strategy, in which the only way one could protect each new frontier was to create another still further out. Where aircraft became abundant, there seemed no limit to the capacity to project fire control. For even the most pyrophobic countries, however, this could not occur until after World War II as governments made available war-surplus machinery, expanded their range of services, and began to harvest the protected woods.

Two problems emerged, however. One was economics. Mechanised fire suppression was costly, and aerial fire protection breathtakingly expensive, even when subsidised by surplus equipment and national treasuries. A point of diminishing returns kicked in, and because big fires only occurred episodically, it was easy, after a few quiet years, to decide the forward policy was unnecessary and to reduce the budget. But this kind of economising voided the strategy, for it was only by having the apparatus for aggressive initial attack, always and everywhere on-hand, that you could forestall conflagrations. Instead, politicians adopted a casino philosophy, in which climate rolled the dice and you took your chances. When the roll was favourable, the monies went elsewhere, and when it went bad, it was easy to justify fresh expenditures during a full-blown emergency. Besides, public lands did not have voting rolls.

The other issue was ecology. It became increasingly apparent that fires not only happened but that biotas expected them to happen and could be upset by their removal, or since adaptation occurred not to fire per se but to fire's regime, they could be unhinged by a shift in the patterning of burns. Philosophically, then, there was an argument for letting natural processes, not least fire — even high-intensity blazes — proceed untrammelled. Scientifically, the case grew that fire exclusion could be damaging, that the core concern was not 'protecting' wild reserves from wild fire but restoring the fires lost through settlement. That included the formal fire-protection programs of state-sponsored forestry. In this reckoning, the unmooring provocation was not bushfire but bushfire protection. The most direct, and economical, way to rectify this imbalance was to stand aside and let nature set the agenda and rehabilitate the wounded landscapes.

It is always easier to criticise what has happened than to anticipate what might occur. It was simple to list the conceptual and practical failures of fire protection; any counterarguments remained rhetorical and hypothetical until they got implemented on the ground. Only then did their liabilities become evident. Lingering fires could smoke for weeks, or over a season, creating concerns with public health. They might not remain within their gazetted preserves, occasionally bolting free, now big enough that it was impossible to drive them back and staggeringly expensive to try. No one knew how to cope with high-intensity fires, which were vital in some places for the biological work they did but intractable in others when they spread beyond their designated places. The usual response was to argue that the reserve had to be enlarged; but that led to a let-burning argument that mirrored the forward-strategy argument of fire suppression.

More seriously, let-burning acquired dilemmas of its own. One was pragmatic. Flame was not ecological pixie dust: it could not, of itself, magically 'restore' a disturbed land to health. Since fire synthesised its surroundings, messed-up forests tended to yield messed-up fires. Particularly if fire had been removed, or the regime upset for any length of time, a 'natural' fire might not burn naturally. Behind this quandary lay a philosophical one, that the putatively natural state might well have resulted from a long interaction with indigenous peoples before the land was reserved. The regime to which the biota had adapted might reflect — probably did — a perhaps ancient tenure by anthropogenic firesticks. Lightning alone could not recapture that lost state. Rather, a natural approach might create conditions never before experienced.

In nature's economy of fire, laissez-faire had its limits. It could work in reemote settings, especially those committed to nature preservation. It could not absolve humanity from the need to choose how and where to intervene.

Suppressing

Fire 'suppression' is a means, the preventing of unwanted fires from starting and their rapid extinguishments when and if they do break out. In this sense, suppression is an abbreviation for fire control and can thus support any larger strategy; it will always be with us. But suppression can also distil a deeper agenda, fire's exclusion. In that configuration, it forces other practices to support its ultimate end.

The fascinating question is: How did reasonable men come to believe that suppression was possible and necessary, and worth committing the power and prestige of the state to enforce? An answer is more complicated than the usual critics of fire suppression, who have seen it in its triumphalist phase, are willing to grant. An answer must absorb environmental idealism, a historical moment, and imperial politics.

Begin with the European origins of colonialism and forestry. Temperate Europe — where, from the 18th century, the main powers of Europe resided; economic, intellectual, military, colonial — had no natural basis for fire. Temperature changes with the seasons, but precipitation remains, month by month, relatively constant. The European landscape ideal is a Garden, with everything (and everyone) in its proper place. Fires broke out during times of unrest, and were set by those in motion such as wanderers, swiddeners, shepherds, soldiers at war, and peasants unmoored by famine, plague, or conflict. The only fires allowed were those on candles, in hearths, or in furnaces, or where used in agriculture, in a well-disciplined order of field rotation. (Agronomists, however, hated fallow, and detested its burning.) As a graft on the rootstock of European agriculture, forestry shared these presumptions. The ideal woods was a kind of orchard: useful, pretty, and green.

What Europe found in the rest of the world — places far more prone to burning — it condemned as unruly and primitive, and in need of rationalisation on the European model. Beyond the Garden, however, colonisers could control little: not the climate, not ignition, not the arrangement of fuels on the landscape. Of all earthly exemplars possible for fire, Europe's became the best known and the least usable outside its formative hearth. Yet foresters carried these ideals in the rucksacks. Ideally they would have liked to raze the unruly wild woods altogether and replace them with plantations, and if that was not possible, as it very rarely was, they would improve the character of the indigenous forests. Eliminating fire belonged with the abolition of plagues and banditry.

Utopia met history, however, during the great colonial surge inland. To the indigenous burning, settlers added land-clearing fires, and on their heels came the pyric transition. By the time forest reserves were being established, many colonial landscapes were a shambles, and fires were intertwined with that wreckage as agent, aftermath, and symbol. They encouraged pastoralism at the expense of woods; they helped clear majestic forests for hard-scrabble farms; they followed the frontier as plagues did armies. Fires swept through slash and burned over villages. The vanguard of industrialisation, notably the railroad, cut new corridors of flame and threw sparks to all sides. If there were good fires, they seemed minor compared to the malicious ones that roamed the countryside. And if there was one item that everyone — both conservationist and developer — could agree on, it was the imperative for fire control. Fire protection was, in fact, a dominant reason for creating reserves in the first place. The task was to rationalise and modernise these landscapes, and protect them from humanity's bad behaviour.

And finally there was the question of who would hold the firestick. Controlling fires was a way to control the local population. They couldn't live in the old way if they couldn't burn. Conversely, arson was a preferred 'weapon of the weak'. In the colonial context, fire control was a measure of political order, and when transferred to gazetted reserves, it was a public expression of power. If the state could not protect its privileged forests from fire, it was visibly enfeebled. In democratic countries like the U.S. and Australia, foresters claimed that they could impose order where pastoralists, prospectors, and shifting cultivators could not. Their capacity to control fire became, by their own declarations, a test of their ability to administer those reserves properly. Compromise seemed unthinkable.

Those claims did not go unchallenged. Early burning, light burning — the critics were as ardent as the foresters. But suppression had its own internal logic and, once begun, it could rack up many successes. Almost everywhere the early achievements seemed impressive. By abolishing local burning and installing a first-order network to detect and attack fires quickly, burned areas plummeted. It happened in Australia, South Africa, North America, and the boreal fringe of Europe. There were failures, of course, but administrators could plausibly argue that these were outliers caused by extraordinary circumstances, whether climatic or social, and that, as suppression strengthened its grip, it could eliminate them or hold even exceptional fires to acceptable limits. The indispensable need was to find and hit those fires, all fires, early.

Fire suppression, however, is not sustainable, to use contemporary jargon. Initial attack could only succeed by making the reserved forests into the environmental equivalent of a police state, and that classic set-piece, the big firefight, proved akin to a declaration of martial law. It was adequate to cope with a temporary insurrection, but it was not a basis for governing. If there were local residents, they were in constant rebellion, and if the lands were uninhabited, the woods themselves simmered with insurgency. It was impossible to prevent all fires; the big fires would become bigger. Instead of steadily imposing a new order, suppression — or, more broadly, fire exclusion — created progressively greater instability.

But the news got worse. Not only did fire exclusion fail to protect fire-prone forests, it was a menace to their biological integrity. Removing fire might boost logging revenue from a leased coupe, but it could very well harm a preserved place that had adapted to a particular regimen of flame. This applied even to big fires because there were biotas that regenerated through immolated crowns. Eliminating those high-intensity fires, which seemed the essence of the project, might unhinge the landscape as surely as sending mobs of sheep through them. It became increasingly unclear where fire suppression should ply its craft, outside the urban fringe, but unless it could contain fires everywhere it could not guarantee protection anywhere.

On almost each particular, the original critics have proven correct. The only way to fight fire, it seems, is with fire.

Controlled burning

If fire there must be, the argument developed that it was wiser to do the burning yourself. Better fires on your terms than on nature's. Better leashed fires than wild fires.

The thesis could mean different things to suppression's critics, however. To foresters, controlled burning meant reducing fuel. It was an inexpensive, benevolent way to improve fire control; it was suppression by indirect means. The manifold adaptations of many species to surface fires assured that the fires would do no unwarranted damage. By co-opting fire into their tool-kits, foresters could accomplish what brute-force suppression could not. Removing fire had made its control more difficult; restoring fire would bring it once more under restraint.

To environmentalists, however, controlled burning was measured not for the boost it gave to 'control' but against its effects on the biota and as an index of political power over the public estate. Here, one might recall Winston Churchill's observation about Britain and America as two people separated by a common language. In this case, America and Australia diverged over what the apparently common language they shared actually meant.

In America, the forcing mechanism was passage of the Wilderness Act (1964), which reclassified some public lands but left them under the jurisdiction of the original agencies. Thus, wilderness areas could exist within national forests, national parks, wildlife refuges, or other public land. Instead of creating a national wilderness service to administer those sites, they remained under the National Park Service, the U.S. Forest Service, the Fish and Wildlife Service, or the Bureau of Land Management. The feud over prescribed burning was a fight over fire management, or more broadly, land management within those special areas, not over jurisdiction for that estate. Although federal foresters resisted reclassification of national forests as wilderness, they nonetheless remained in charge of the lands that Congress so designated. And although they first protested that wilderness classification would prevent them from controlling fire, the reforms sparked by the wilderness movement became the basis for shifting strategy from fire control to controlled burning. Paradoxically, the identification of free-burning fire with the wild boosted the appeal of prescribed fire, which could be seen as spreading the natural goodness of the wild to other places.

Environmentalists' enthusiasm for restored fire was partly ideological and partly practical. Over the past century, America's public domain had gone from a fire-flushed land to a fire-starved one. The outcome was widely regarded as an ecological shambles, and the hope grew that reinstating fire would reverse that degradation. The preferred means was to let nature's fires do nature's work; but granted the profound disturbances wrought by fire's exclusion, environmentalists accepted that deliberate burning might first be necessary. (Interestingly, this mirrored the 'prime directive' promulgated by *Star Trek*, which began its TV run at the same time — non-interference; but if interference occurred, then one was obligated to restore the prior conditions before withdrawing.) The National Park Service converted to the new policy in 1967–68; the U.S. Forest Service, after some temporising measures, followed suit in 1978. In America the philosophical fight over fire doctrine ended decades ago.

Australia differed. A reclassification of lands brought a transfer of jurisdiction to another agency; wholesale transfers could mean the effective extinction of state forestry bureaus. At issue was not simply professional identities and doctrines of land management but administrative control over public lands and the power that goes with that control; the stakes were higher, the quarrel more bitter. Complicating the controversy even further was the fact that Australian foresters had already committed to a doctrine of widespread burning. The American critique of foresters as mindless fire-fighters could not apply: Australia's foresters were ardent burners, certainly in principle. Paradoxically, the critique turned on the fact that they did burn. Critics sought a more 'natural' fire, namely, from lightning. They dismissed forestry's hazard-reduction burning as ecologically malign because it did not obey such a rhythm, and if followed with metronomic rigour might actually homogenise the biota rather than promote such valued characteristics as biodiversity. Fire protection through hazard-reduction burning would destroy biotas more surely than wildfires. Because of their fire practices, foresters could not be trusted to administer Australia's most precious lands. Whatever good their fires might do, they were still contaminated by association with practices that metropolitan (and cosmopolitan) Australia disliked. The criticism foresters had traditionally directed toward forest graziers was now directed to them. Associated agendas intertwined with fire to make a political briar patch.

But did controlled burning of any kind belong? It was one

thing to dismantle the case for hazard-reduction burning, which existed to improve fire control; it was another to argue for an ecological analogue. In principle, a green agenda could tolerate limited fuel-reduction burns as well as naturally kindled fires, and a case could be made that some deliberate burning might advance the cause of select species and habitats. In practice, greenies, like imperial foresters before them, saw a landscape overrun with fires they didn't like. There was simply too much; and the beloved precautionary principle argued for snuffing out as much as possible, doing the proper research, and then, if and as needed, grasping a gentler, greener firestick.

Fire policy thus became a proxy for the whole shebang of environmental politics. State forestry argued that it was the repository of fire experience, and that because it knew fire, it probably knew best how to steward the land overall. Environmentalists argued that they were the repository of ecological understanding, and because they knew best how to care for the land, they knew best what to do about fire. Specifically, they often denounced forestry's burning as unneeded, unwise, and ecologically unwelcome.

There was a wild card in the deck, however. How should one consider the ancient tenure of Australia's Aborigines? Had their multi-millennia presence existed for so long that they and their fires had become, in effect, naturalised? Was the regime to which Australia's fabulous biota adapted one that included the Aboriginal firestick, and that in the absence of that firestick would disintegrate? If so, there was a case for a human presence in nature reserves. More worryingly, there might be a case for widespread anthropogenic burning, which could, to a confused public, look suspiciously like what foresters were doing. After all, foresters claimed their hazard-reduction burning was but a modern avatar of the Aboriginal firestick, only carried by aircraft rather than by hand. The intellectual justification for and against burning was beginning to look like real work.

Here again, Australia and America diverged. By definition, America's wilderness precluded a permanent human presence, which had only arrived in any event during the long upheaval of the Holocene following the continental glaciers; Australia's preserves could hardly ignore 40,000–60,000 years of constant human habitation. With the issue unresolved, America's administrators were nonetheless committed to reintroduce fire. Australia's were determined to reduce traditional burning. One task proved as formidable as the other.

Americans discovered that fire, once removed, was not easily restored. The land changed, lore was lost, and new circumstances shackled the opportunism that had made landscape-scale burning work, patch by patch, season by season. Rather, the reintroduction of fire more resembled the reintroduction of a lost species, say, a wolf. The project would succeed or fail according to whether the proper habitat existed, which was especially the case with fire since it had no other identity than what it synthesised from its surroundings; unlike a wolf, it could not lope elsewhere to a more supportive setting. There were successes — the acres burned mounted — but the cost to fire's use threatened to become as exorbitant as that of fire's suppression. (Between 1995 and 2000, the burned area on federal lands more than doubled, but costs leaped five-fold.)[24] And there were plenty of failures —

not only fires that blew up, but fires that fizzled without doing the biological work demanded.

Australians, meanwhile, found that removing the bad burning did not ensure that the good burning got done. Under the new regime, the amount of land burned overall declined. This helped those sites that needed a rest from flame, but did nothing to quench a surface build-up of combustibles where the land demanded protection or to provide flame of the right sort for places that craved it. Foresters had argued that their patterns of fuel reduction burning had been biologically benign, and had sustained habitats in the process. Now environmentalists had to argue that their regimen of selective burning, and particularly their regimen of not burning, could support a necessary level of protection. The final balance sheets could be eerily similar.

Except where the bureaucratic bottom line applied in the form of funding. Forestry had paid for itself by revenues from logging and leasing, and had justified its emergency fire expenses by its labours to protect lives, property, and natural resources. Environmentalism had come to power with the implicit assumption that by doing less it would cost less, that the protection of biodiversity would not demand panicky outlays of public funds during crises, that the new order could protect by withdrawing. Proponents for the Australian strategy considered this an expression of pyric appeasement. Bushfires would not cease because greenies wished well. Worse, it was awkward for green officialdom to argue that the old costs might have to continue in new guises while lacking a comparable revenue stream to pay for them.

For both countries, the lesson might be that putting fire back to its historic level or removing it from those levels would not be easy, cheap, or safe. The one would not be the simple reversal of the other, as state-sponsored ecology replaced state-sponsored forestry. It was not enough to disarm firestick forestry; the new order needed a version of firestick ecology.

Changing combustibility

The fourth option is to alter the properties of the landscape. Modify the arrangement of the vegetative cover and you modify the fire, any fire, that might burn through it — whether begun by lightning, arson, accident, or professional firestick.

This, of course, is the classic European solution, the application of its social model to the environment. A close-tended landscape, a garden, will not suffer wildfire. If it needs fire — say, a burning of the fallow — those flames will remain within the confines of the plot at the time allotted. This reasoning also underwrote some of the thinking about settlement fires: that they were an artefact of land-clearing; and that, once wild bush was domesticated into a pastoral or plowed scene, those fires would vanish. Wild fire was an expression of wild land. The converse was that excluding fire from a place that had too much of it would allow the land to change in ways that would further exclude fire.

This sometimes happened. The problem was that, in the name of conservation, large swathes of country had been set aside and would not be subject to this transformation, and in

Australia particularly, it was impossible to fence off fires for the many decades that the proposed experiment in fire-immunity would demand. The point of the reserves, after all, was to spare them from permanent conversion. But retaining immense wildlands meant that one also retained an immense, permanent domain for wildland fire.

Recently, the issue has been revisited. The immediate provocation is that the urban bush is aflame. The agricultural expansion that had caused such friction with the woods a century ago has been replaced by an exurban expansion that may prove every bit as intractable and deadly. Agriculturalists cleared and converted; they were vulnerable to horrific fires mostly during that time of transition. Exurbanites do not clear their lots, and typically they only half-convert — which is to say, they tend to freeze that transitional phase. Where development broadens, the transition to urban landscapes can lessen the threat from fire. More commonly, these exurban enclaves simply slam the wild and the urban together, the matter and anti-matter of modern environments. A few explosions should surprise no one.

One obvious response is to clear around those enclaves. Reducing and rearranging the vegetation can lessen the intensity of a fire. But 'defensible space' can go beyond the immediate lot line. What works around houses should also work around communities, and should work, too, around and within places being protected for their ecological values.

This was the assumption behind America's Healthy Forests Act and National Fire Plan. The belief had grown, however, notably among foresters and federal land administrators, that restored fire alone might not be able to do the job, that what was needed was a modern version of slashing and burning. Only fire that follows a deliberately altered scene can reduce fuel loads and shield the land (and houses) from a high-intensity conflagration. And if the scheme succeeds around the urban fringe, it might underwrite, at least conceptually, a wholesale rehabilitation of the national estate. The Forest Service estimates that some 43 million acres are seriously out of synch with their appropriate fire regime.

The latter possibility is what alarms environmentalists. What the forestry and fire community call 'fuel' looks to most people like grass, shrubs, and trees, the living habitat that they have collectively determined to preserve. Having shut down logging in most American national forests, they are unwilling to have the industry return under the guise of 'thinning' for fire protection. Better wild fire than wild axes. They view suspiciously a renewed alliance of chainsaw and firestick, and they scorn the rebuttal that the program has as its ultimate ambition the creation of more ecologically stable woodlands. They've heard this before. They trusted 'professional' forestry once and found it wanting; they are not eager to go another round.

The more expansive agenda has not yet entered Australia, where the fight with forestry remains over hazard-reduction burning, in good measure because foresters have insisted that broad-area burning, done right, will lessen the need for mechanical supplements. Australia's foresters believe, based on bitter experience, that the sardonic eucalypt can survive a harder hit by fire than can ponderosa and longleaf pines, the poster-children for the American strategy, and are thus inclined to rely on fire as a corrective. They believe that, while the window is closing, Australia has avoided an American-style crisis because it kept fire on the ground, never in anything like the scale required but in sufficient quantities. If New Australia wrenches that favourable state into something like American conditions, it will face an American-like crisis, an overgrown countryside subject to diseases, insects, and conflagrations, a bulimic landscape full of fire binges and fire purges.

But what both countries share — what seems indelibly obvious everywhere such conditions prevail — is that no single strategy can be the basis for fire's management. There are places where fire can be allowed some elbow room and even permitted to roam — a bit. There are circumstances where suppression is the only option. There are sites where fire will never do what we hope without some prior treatment that allows it to behave as we wish. And there are times and places where controlled burning is not only useful but obligatory.

The American fire community knows that it cannot cut its way out of the problem, or burn its way out; it can't suppress; and it can't walk away. It must find proper alloys of those techniques, adapted to particular sites. Bushfire is, after all, a very empirical creature. America has learned that fire management is not simple or safe, and it is certainly not cheap, and that if one chooses to replace humanity's fidgety hands, you must substitute a fidgety mind. If fire management cannot be labour-intensive, it must be information-intensive.

That will likely hold for Australia as well. Probably, Australia will have to invest as seriously in its management of fire as it has for water. What seems true for Canada, Russia, and America seems destined to prove true for Australia as well. With this difference: the preserved fire, the legacy of firestick forestry, will make the task easier. Australia held on to a firestick that the others blew out and are trying to rekindle amid the ecological darkness that resulted.

The green in the ash

America's 2000 fire season recapitulated in scale and media attention the epic Big Blowup of 1910, and helped galvanise a political reformation, the National Fire Plan. This time the replay of the light-burning controversy decided for the other side: there was no dissent that America needed to get more of the right kind of fire into its public domain, although it was not clear whether prescribed burning alone could do that, or whether it needed the additional leverage of the axe.

So, too, did Australia's 2003 season seem to revisit the formative Black Friday conflagration of 1939. Even before the Alpine fires broke out, the comparison was on every lip, and has shaped the discourse that followed. This time foresters assumed the role graziers had in 1939, and indeed found themselves in an improbable alliance with their ancient, blood-feud rivals, saw their extensive use of controlled burning condemned, and witnessed the final act of a regime change in the administration of Crown land. What most differentiated the two episodes, however, was that no royal commission inquired into the 2003 conflagration, and hence no Leonard Stretton eloquently inscribed its lessons. The heroic era ended not with a new Homer but with what seemed to critics a bureaucratic belch.

The reports flow on; there seems no end to them. The primary vehicle was the national inquiry sponsored by the Council of Australian Governments (COAG). Chaired by a familiar triumvirate of professions — a representative of the emergency services (Stuart Ellis), a forester (Peter Kanowski), and an ecologist (Robert Whelan) — it surveyed the country comprehensively, documenting some 54 million burned hectares, summarised the legal and political apparatus for bushfire control, and proposed 29 recommendations. But mostly it confirmed the twin thrusts of bushfire politics since Ash Wednesday: a paradigm of risk management to cope with wild fire, and a paradigm of biodiversity protection to regulate controlled fire. Both strategies had had national cornerstones fitted a decade earlier with a Commonwealth policy on Ecologically Sustainable Forest Management (1992) and the creation of the Australasian Fire Authorities Council (1993).

All in all, it was a measured document, a grand expression of political technocracy, whose very reasonableness denied it cathartic power and thus pushed it to the margins of public controversy. Instead, the wrangling gathered around two inquiries: one sponsored by Victoria, which carried the voice of the new order; and the other, a coronial inquest in Canberra, in which the critics hoped for a bolder hearing. Revealingly, the coronial inquiry under Maria Doogan has faced a long slog, culminating in a suspension in October 2004 on charges emanating from Jon Stanhope, chief minister of the ACT, of 'apprehended bias' (in favour of forestry-based critics). Not until August 2005 did the Supreme Court decide that it might proceed, dismissing protests as 'premature' and concerned with 'circumstances that have not arisen and might never arise.'[25]

The inquest had, in fact, been hounded by confusions, partly over the exact scope of the proceedings (whether into the 'cause and origin' of a fire or of a disaster), and especially by the role of N.P. Cheney as both investigator for the coroner and an expert witness who had, in the early stages, accompanied Coroner Doogan on a field trip to relevant scenes and offered extensive comments. Further stirring the stew, Cheney, a government critic, was also a government employee, completing 30 years of bushfire research with CSIRO. Alarmingly for those defending ACT officials, he was an unapologetic and outspoken protégé of A.G. McArthur, and one widely recognised internationally as a voice of Australian fire. All this led attorneys defending ACT officials to fear that the coronial hearing, as the only quasi-judicial venue extant, might become in practice if not in name the royal commission that the Establishment had categorically denied critics.

While the coronial hearings dragged on, Victoria rapidly completed its *Report of the Inquiry into the 2002–2003 Victorian Bushfires* (generally known as the Esplin Report, after its chair, Bruce Esplin). The document confirmed the political triumph of the new order. It refused to characterise the burns as a 'disaster', since (unlike its predecessors) this report was 'not inquiring into a major disaster in terms of deaths, injuries and homes and property lost'; nor did it find evidence of 'major systemic failure'. Hence there was no need for a full-scale judicial review. It was sufficient that improvements proceed within the structures already laid down. And while 'community expectations of government and its emergency services' had risen, the board of inquiry, headed by the Commissioner of Emergency Services, assessed the existing apparatus as 'equal to "world's best practice".'[26]

Strikingly, there were no foresters on the board, which included Bruce Esplin, Malcolm Gill, a fire ecologist retired from CSIRO, and Neal Enright, a professor of anthropology at the University of Melbourne. So, likewise, forestry had no institutional role. The grand alliance it had forged between state institutions, rural residents, formal science, and conservation dissolved. Fire suppression had gone to emergency services bureaus; fire science, to universities and a Bushfire Cooperative Research Centre; land management, to parks and preserves; prescribed fire, to those deeply suspicious of it. A generation of foresters — the apostles of McArthur, the second generation of the Australian strategy, now retiring — was outraged. They demanded a stronger voice, and if their preferred venue, a royal commission, was denied them, they would seek it through coronial inquests, counter-reports, and the formation of non-profit organisations that would allow them standing akin to that accorded environmental NGOs. They objected, not least, to the smug tenor of the report, which to their ears had an apprehended bias far worse than anything manifest in the coronial inquest.

Providentially, the 2003 Alpine fires had started mostly from lightning, not 'the hand of man,' which made them appear more a force of nature than a social meltdown that another, better disciplined group of managers could correct. This, so official reviewers insisted, was a natural event, no more controllable than Cyclone Tracy or the Long Drought that closed the 19th century. Moreover, such fires were what the mountain parks required. Like most Australian plants, mountain ash was a phoenix flora, even if its cycle of fiery regeneration spanned centuries. Besides, even 'conflagrations' burned with patchy intensities that left a mosaic of variously scorched land. There were breakdowns once the fires spilled out of park borders, and naturally there were improvements possible in the composition and efficiency of the emergency response; these were inevitable. But the board stoutly denied that a 'disaster' in the conventional sense had occurred. What the Alpine region had experienced was a tantrum of nature, the ecological equivalent of an earthquake.

Critics disagreed. Forest Fire Victoria, a non-profit group of mostly retired fire specialists, argued that many of the fires could have been dampened (and maybe extinguished) had they been attacked aggressively during a nine-day lull that followed ignition. A cognate outbreak of lightning fires, under broadly similar circumstances in 1985, had been hit and held to far smaller dimensions, but that was when state forestry still had the muscle to do the job and before the doctrine and machinery of prescribed burning had been unravelled. The 2003 fires had blown up because the crews, machines, tracks, and perhaps the will to overwhelm the flames had eroded along with state-sponsored forestry, and the commitment to fire protection had been hollowed out by fifth-column critics and fire ecology's conscientious objectors. This was overtly a criticism directed to the emergency response, the fire-fighting. Behind it lay a deeper critique that a reorganisation of land agencies was in order; that the existing arrangement had not done sufficient burning to help leverage the suppression forces that did exist.[27]

In fact, the recommended level of burning had never been

achieved. In his day, Stretton had ridiculed the amount executed; after the 1983 Ash Wednesday burns, an inquiry repeated the charge that, even with the Australian strategy in full flush, the agencies were getting only a fraction of the requisite flame onto the ground. Statistically a large chunk of what had passed as 'prescribed burning' had been 'regeneration burns,' fires in the slash left by logging that both diminished their potential for feeding a wild fire and promoted new growth. While this certainly reduced hazard, it was a hazard that forestry had itself caused. The lines-of-fire and fields-of-fire that a serious commitment to fire management demanded were withering rather than putting out new shoots.

In an eerie way, the continued controversy again reversed the position of state foresters vis-à-vis the Stretton Commission. If their land practices placed them with graziers, their inability to do the burning that they insisted was essential aligned them with the grandees of European forestry like C.E. Lane Poole. No, Lane Poole had confessed, his staff had never excluded fire long enough to test the argument that a landscape once spared fire for a sufficient period of time would continue to exclude fire spontaneously. Stretton replied that they never would. No, the Australian strategists might admit, they had never succeeded in doing hazard-reduction burning on the scale needed. A critic might retort that they never would.

Nonetheless, controlled burning would not go away. The board's professed agnosticism — that much more research needed to be done — critics dismissed as a de facto stall. There would never be enough research; never. The statistics, as always, could be mustered by either side to suit its theses. Statistics were as thick as fireweed. Data flowed like post-burn runoff, and proved almost as murky. For every anecdote, a counter-anecdote existed; for every critical example, a counter-example; for every hard-bitten fact, another proving exactly the opposite; or rather, that proved nothing. Like different salts stirred into a vat, the combined numbers seemed to dissolve. The issue would not, in truth, be decided by numbers, or even by science. Resolution lay with the larger politics of what the Australian electorate wanted its Crown lands to be. The tables, figures, graphs, and anecdotal numbers that salted the yeasty mound of documents were so many vignettes decorating the margins of that text. There was thesis and antithesis — but the synthesis would come from a metropolitan electorate.

The appeal to burning 'smarter' rather than burning more, critics also scorned as a bureaucratic sleight of hand. The Esplin report accepted that there were 'substantial areas' where prescribed burning was potentially suitable, and perhaps 'the only cost-effective means' to dampen fuels. But determining those exact sites and the appropriate fires required ... more research; much more. Until then the precautionary principle favoured not only a light hand but a hand without a light. And despite intellectual arguments, 'ecological' burning was not replacing 'hazard-reduction' burning; probably less than 3 per cent of overall burning had ecological goals. Firestick ecology had not replaced firestick forestry: the firestick threatened to go out altogether, which would lead to even greater dependence on robust emergency services. The public-health analogy was that a diminution of preventative measures — vaccinations, proper diet, and the like — would force more effort into emergency response, which had the additional political virtue that it was visible and dramatic. Of course Australia needed more smart burning; it also needed a lot more good burning, full stop.[28]

Layer by layer, its critics peeled away the veneers that, together, had bonded into the Australian strategy. Its architects had pointed to the manifold adaptations to fire by the dominant flora, notably eucalypts, and had argued that burning of the right sort would do them no harm. Critics discovered other species, and mosaics of species, often in niches, that might be damaged by a regular regimen of low-intensity winter burning. The founders of the Australian strategy had noted the erstwhile presence of the Aboriginal firestick, and suggested their own burning perpetuated that tradition. Critics demanded impossibly precise accounts of site-specific burning by Aborigines before they would accept an ecological lineage; and surely, Aborigines had not rained incendiary capsules from the sky. They argued further that 'prescribed burning 'should be the preferred term, dismissing 'fuel-reduction burning' as a 'tautology' and 'hazard-reduction burning' because it 'begs the question of what the "hazard" might be'.[29]

Both sides could make a straw man of the other. Conservationists could burlesque hazard-reduction burners as clear-fellers by another name, determined to force fire, on their rotational schedule, down the throat of every landscape in Australia. Foresters and their rural allies could caricature environmentalists as more concerned with an obscure species of nocturnal marsupial than with country folk, and as wilfully keeping proper fire out when that only meant a worse fire would come later. Again, the furore was no more about knowledge than about numbers. No one had perfect knowledge; everyone could easily criticise the claims of others, quite apart from any phoney reconstructions of their assertions. The uproar was about cultural values.

That is what made the discussion intractable. It forced participants to align with professional, social, and political identities. Fire practices were code names for groups, for the larger values and practices of those groups, and for the political stature those groups claimed. There was truth to that fear that fire policy was a means to other agendas. Graziers might use their heritage of burning to argue for a reintroduction of alpine livestock, which unquestionably damaged soil and flora. Foresters might imply that, being right on fire, they were also right on other matters of land use and sneak a bit of logging under the radar screen. Environmentalists exploited the tainted association of fire with grazing and logging to call any fire use into question. They dismissed the accumulated lore of the firestick foresters, as European foresters had discounted the practical knowledge of farmers and graziers, and as settlers had often ignored the fire wisdom of the Aborigines.

Behind the rants and submissions and expert testimonies lay that deeper query into identity. Greenies had no clear reply to the oft-brusque rural nationalism of the Australian strategy, save a bleached cosmopolitanism that could mingle Erickson aircranes with the fire-threatened habitats of pygmy possums. Both sides might ally themselves with the mythical Bush, but one saw the bush of pioneers and the other, a bush untrammelled by human habitation. Was Australia the land or its people? Was Australia's formidable nature something one simply had to endure, or were ameliorations possible? Was the country best served by picking up the firestick and redirecting it to new purposes, or by finally discarding that implement as an anachronism in the human toolkit that should have been scrapped millennia ago and had somehow lingered in an island continent forgotten by the rest of the world? That question no amount of research could resolve. It lay in the hearts and minds of Australians and could only be settled by politics. The Esplin report confirmed that, for the moment at least, the politics had been decided.

PART IV

Pyromancy: divining futures in the flames

What might the future hold? A few trends are evident: some apparent globally, others peculiar to Australia. They will likely persist. As that noted folk philosopher Damon Runyon put it, the race is not always to the swift or victory to the strong, but that's where you put your money.

The big burn

The master narrative remains the silent one: the spread and intensification of industrial combustion, the new prime mover of fire on Earth. Here is the real Big Burn; here, the real disturbance in the Force. Nature reserves and the urban sensibilities that inform them are the products of industrialised societies. The removal of land from agricultural pursuits (and the abolition of their associated fire practices) is something developed countries can afford. The prospect for fire-fighting by aircraft, motorised pump, and rapidly transported crews is possible only through the power of internal combustion. Even the climate itself is being reworked by the greenhouse-gas effluent from the combustion of fossil biomass. The Earth continues to segregate into two grand realms.

The prospect is accelerating. There seem to be stocks of fossil fuels ample to last for centuries, and especially as China and India make the industrial transition, the amount of anthropogenic combustion will metastasise. For a few countries, with stubbornly rooted rural villages, the transition will be halting, and traditional burning will yield slowly; the land will become a pyric palimpsest. The plague of eruptive fires over the past decade is, in this sense, illusory: the dominant future is not one of resurgent fire but of receding fire. One could, in fact, argue that many of those explosive burns were a side-effect of industrial land use, that they occurred where lands had been clear-felled and littered with slash, where peatlands had been drained and dried, or where land had been left in a quasi-natural state and subjected to some sustained period of fire exclusion, which ensured the mandatory fuels were in place when drought, wind, and spark converged.

At any place one realm of combustion or the other tends to dominate, save for a period of transition. These are local landscapes altered by a society either powered by fossil fuels or linked to an industrial economy through transport. Over all these proliferating points — integrating them — wraps the atmosphere, where the effluent of lithic landscapes is concentrating and refashioning a common climate. Combine local effects and global causes, and you have the makings of a planetary fire age. Australia again found itself within a larger fire agenda, this time not as part of the British empire but of the Pyrocene. Ark Australia cannot avoid these changes: the atmosphere is gathering the by-products of all and every source. A climate that it has recognised has never been its to control will become ever more beyond its reach.

How these changes affect bushfire is tricky to forecast. What matters is not global warming or cooling as such but the reconstituted rhythms of drying and wetting. Two general scenarios float above the rest. One suggests that global warming will push the overall level of burning (at least in regions) to a generally higher level, that the fires-on-steroids are here to stay. The other proposes a violent period of transition, characterised by spectacular burns, out of which a different biota is birthed, complete with its own suite of fire regimes. But while both point to climate change as a driver or forcer, they fail to appreciate the link between the firestick and the combustion chamber, that humanity's fire habits are powering the changing climate. The driver is not climate but humanity in our singular capacity as a fire creature.

The industrial transition will complicate the use of open fire. Most of the clashes will concentrate in the atmosphere, where free-burning fire and industrial combustion compete for limited airshed. There will be concerns over public health from smoke and toxic effluents; aesthetic objections to obscured views; and political calculations about carbon accounting, as global conventions seek to reduce the release of greenhouse gases. Apart from technical difficulties, which can make prescribed burning arduous, even parlous, the deeper drivers of fire history may well prejudice its use.

Public lands, public fires

The second trend concerns the future of public lands. A colonial era established them; a post-colonial era is restructuring them, along with other relics of the imperial experiment. They are being devolved, privatised, disestablished, hollowed out, returned to indigenous peoples, subjected to community-based conservation, reclassified for new purposes, overseen by consortia of which the government is only one partner, or any combination of the above. In particular, large chunks of what had once been forest reserves, or lands subject to timber or pastoral leasing, have been rechartered as parks, wilderness, wildlife refuges, or monuments.

While there is plenty of combustible rural land yet in Australia, especially pastoral, fire protection on them means controlling unwanted burns, and the methods are of a piece with other practices that aim to produce meat, wool, grain, timber, and like commodities; among these is a reliance on agricultural burning, along with organised fire control. The state has an interest as a matter of public safety, but the local community may self-organise, as Australia's have so often done with volunteer bushfire brigades. As the pyric transition persists, however, the pressures will grow to replace that open flame with alternatives. If those lands are to remain competitive, they must eventually intensify their husbandry, which will tend to squeeze flame from the scene. If not, then they may become part of a fractal urban fringe, or revert to Crown land as leases expire.

The real issue resides within the public lands. These have

become the prime habitat for free-burning fire and one of two flash points for contemporary fire 'problems' as defined politically (the other being the bush-encrusted urban periphery). Because they remain as bush, they retain the potential for bush fire, along with the requirement (perhaps) for some degree of burning. They involve the state directly: the government has a duty of care to the lands within its estate. But as those lands acquire new purposes, so they must fashion new institutions for the administration of fire. The programs and policies devised by the founding foresters may no longer suit. The transition period, unsurprisingly, will be one of uncertainty, perhaps fumbling, as the public struggles to define what it wants from those lands and as administrators puzzle over practical ways to make those ambitions happen on the ground.

Fire follows fuel. Changes in land use on the public domain will release or sequester combustibles, which will determine what fires are possible. Changes in policy and institutions will affect how people shape those possibilities. There will be both more fire and less, and different fires. As vast chunks of Australia alter their character, fire will change with them.

The colonisation of new new Australia

Like other industrial countries, Australia is finding its rural landscapes recolonised by an urban out-migration. The metropolitan-based New Australia that emerged in the post-war era is morphing into an exurban New New Australia. Like the earlier wave of agricultural settlement, the frontier has become one of damaging fires, even as suburban enclaves rather than farms, and parks rather than paddocks, remake the countryside. Previously, the bushfires had generally resulted from slashing and burning, or browsing and burning. Now, the fuel-laden landscape results from not removing the vegetation and not burning.

This is a dumb problem to have because technical solutions exist, as they don't in nature reserves. These are exurban enclaves; we know how to shield houses and homeowners from fire. But in shaping a measured response, we might consider the symmetries between the two eras. Edicts, bans, disdain, threats and force, fulminations against the very tide of settlement — these proved relatively worthless, however gratifying as a rhetorical purgative. Rather, government worked best when it joined with the local agriculturalists, when it helped them do the necessary burning, assisted them in organising bushfire brigades, advised settlers and directed them to better, less dangerous sites. So, today, ridicule and threats seem unlikely to stem the outflow. What government can do is bolster protection against wild fire, assist with sheltering homes and organising community responses, and try to channel the demographic flow away from the poorer, more hazardous locales.

All in all, Australia has done well with these themes. Some of the earliest and best research on the question of why and how houses actually burn emanated from the wreckage of the Ash Wednesday fires. That tradition has continued. Where the rest of the world has plunged into the topic, they are generally confirming the Australian findings. The project has extended into how residents, not merely houses, should survive; whether to flee or fight, and how to prepare for either. Again, these studies, and their embodiment in such programs as Victoria's Community Fireguards, have pioneered an approach that has caught the attention of researchers elsewhere.

What seems lacking is the perception that the urban bush is not the outcome of individual ignorance and class obstreperousness but of a larger social enterprise, common to North America, Europe, and such patches of industrial modernity as South Africa's western Cape. Eventually the demographic tide will ebb, though not for a good while yet. What seems lacking, too, is an expansion of understood obligation. At least in principle, homeowners accept the notion that responsibility for defending their home resides with them, from the need to install incombustible roofing to doing what they can to protect against ember blizzards. No similar conviction has collectively leapt from the urban bush to the deep bush, which is to say, has prompted a comparable recognition that they must extend their housekeeping from the backyard to the back country.

The flickering firestick

The question of how much fire to put in and take out, and by what means and to what purposes, haunts public lands everywhere. In Australia this will likely mean that the controversy over controlled burning will endure, under whatever avatar and label it acquires.

The real issue is not the debate, which will surely persist, as arguments grapple with one another and data pile like slash. The real issue is whether the firestick makes contact with the land. The pressures against open burning will mostly mount; and arguments to replace hazard-reduction burning with ecological burning do not guarantee that it will happen. Left to laissez-faire, it likely will not. Concern over improper burning will add another voice to the chorus objecting to burning of any kind, and further encumber the opportunism and agility that a deft program demands. There will be justifications to delay, to research further, to thicken the stipulated conditions that must be satisfied before lighting up. The precautionary principle will urge that the firestick should be placed safely to the side until all the concerns are worked through. There it will eventually burn out.

Yet Australia is different. Here, the precautionary principle would seem to argue for keeping fire on land, and the firestick in the hands of people, while the pondering and the research and the queries flare and smoulder. Once lost from the land, fire can be tricky to restore. That it transferred the firestick to forestry in the post-war period granted to Australia immense leverage lost to other developed nations. What is needed are now institutional and intellectual catalysts to ensure the transfer occurs once again, that firestick forestry reconciles with firestick ecology.

Nature reserves must recognise that fire does biological work that nothing else does, and that their task is not simply to protect against the bad burns but to promote the good ones. This will cost money and staff; fire management is not fire-protection lite or ecology on the cheap. As administrators segue out of a commodity-economy of fire into a service economy, they can remove bulldozers and chainsaws but only if they replace labour-intensive field work with information-intensive programs, which are, if in a different way, also labour-intensive. Down-sizing suppression forces will mean upgrading the staff for prescribed fire and fire research. Dampening the exorbitant costs of once-a-decade mammoth fires will see funds funnelled into higher annual expenses for environmental monitoring. There is, in brief, little reason to believe that fire's management will ultimately be less expensive than fire's suppression. For one thing, suppression will still be necessary; for another, information-based power only shifts the costs. Whatever configuration Australia adopts, fire won't go away, and neither will the outlays for administering both its application and its removal.

To date, the revolution in fire institutions has followed a revolution in values, but to secure those reforms will require a revolution in ideas as well. The best prospect is to reconceptualise fire as a biological phenomenon. The core thesis is simple. Fire is a creation of life: life provides the oxygen; life furnishes the fuel; and the arrangement of combustible biomass, set by evolutionary and ecological processes, determines the character of the burning. As a synthesiser of its surroundings, fire necessarily takes on the traits, the tempo and nuances, of the living world without which it cannot occur. Traditionally, fire has been envisioned as a physical phenomenon that affects biotas, as through it were a heat-flux equivalent to a flood or a windstorm. Fire ecologists assess the character of the landscape before the fire, and then after, examining how the biota survived a fire that smashes into it, as though it were a house. The perception that fire is a physical event suggests that the proper response should be physical countermeasures, that fire's heat be met with quenching agents, that its fuels be physically rearranged, that a breakout fire is an exogenous crisis that requires an emergency deployment similar to other community disasters and can be 'mitigated' by similar measures.

A biological conception would imagine fire less as a physical process that slams into the living world than as a quasi-organic process propagated within and by a living substrate which manifests itself by physical means, notably heat and light. It more resembles an instantaneous epidemic, a contagion of combustion, like an ecological SARS, than a mudslide or a hurricane. Instead of physical counterforces, one would search for biological controls, a model that might look toward the ecological equivalent of vaccinations and quarantines, and landscape patchiness that would integrate those browsers and decomposers that compete with flame for the small particles. Instead of an emergency, many fires might be tolerated as an inextinguishable background count of low-intensity infections like winter colds. Prescribed fire might resemble an annual flu shot — not a preventative vaccine, not always effective, most critical for vulnerable populations, but far better than doing nothing.

The range of options not only expands, but maps onto the language, ecological concepts, and philosophy that lie behind nature reserves. In this context, biodiversity does not confront a physical threat: it must absorb a process that the biosphere itself sustains and shapes in ways both tiny and huge. Instead of reducing the landscape to carbon bullion, ready for slashing and carting into more favourable arrangements, one can consider a more subtle ecological engineering that would aspire to get the right fire in the right place. The problem becomes not to reduce fuel, and hence protect assets from fire, but to orchestrate the proper arrangement of biomass to get the kinds of fires desired. That will, of necessity, mean that the 'fuel' exists in suitable amounts and forms. But this discussion occurs within a biocentric framework. Many places would find that the imperatives for biological burning would argue for far more burning than hazard reduction ever could.

Not least, a biological conception of fire would allow a place for ourselves around the flames. Only when a creature acquired the capacity to kindle and apply fire did the living world complete, in principle, a fully biological control over flame. The ecological value of such a creature would be obvious were it not ourselves. A truly biological understanding of fire would insist on our presence. There may be good reasons for us not to do many things in the natural world, but there is no compelling argument not to be the fire creature we are. A genuinely biocentric philosophy would insist upon it.

The still-burning bush

What should Australia do about its bushfires? That is a matter for Australians to decide.

Australians have a finely honed and most generous habit of hosting visiting luminaries and letting them hold forth. Then they ignore that impassioned rhetoric to do what they must do or want to do. Especially in a matter like fire, which is nothing if not a study in context, the particulars matter. The details of site, of dry spells and wind, of grass and scrub, of institutions and understanding — these are what fire synthesises and what any attempt to manage fire must comprehend. No outsider can master the requisite detail.

Yet that very immersion can make perspective more difficult. What a critical outsider like myself can add to the conversation is precisely the requisite distancing, a conceptual filtering, a geographic, historical, and political place-setting based on comparison and contrast. Australians don't need rants and harangues about the minutiae of fire's ecology or who they should be. They do need to know that their problems are shared widely; and that, if Australia has much to learn, it also has much to tell the others. In that spirit, and instead of prescriptions, let me offer a few observations.

One: In the realm of bushfires, politics is not an intervention or a distortion. It is the fundamental arena for deciding what should be done and how to do it. Bushfire management involves questions of values, of cultural choices, of social arrangements and investments. Especially when the action occurs on public lands overseen by public institutions, the public alone can decide, and politics is the appropriate venue. Technology can only enable, and science can only advise; neither can decide. When technocrats or scientifically informed bureaus have aggrandised that job to themselves in the name of depoliticising it, the outcomes have failed because they were in truth doing politics by another name. The issue is not whether politics should be present, but that the politics be fair, informed, and open.

Two: Celebrate the heritage of the firestick rather than treat it as an embarrassment. America and Australia are today the premier places for bushfire management, where the science, the political interest, and the controversies are keenest. Appreciate Australia's contributions to the global saga of fire; its science and its art, its bushfire brigades, its controlled burning. Australia is a global firepower: it should act as such. Here, in particular, is a place where Australia might direct a significant fraction of its national scholarship and where it might devise technologies and practices that the developing world can exploit as it begins fire's industrial transition. Countries like Brazil, for example, don't need European fire engines or American-style hotshot crews; they need rural fire brigades like Australia's famous volunteers. The developed world could use an alternative to mass evacuations when the flame crests the local ridgeline; something that looks like Victoria's Community Fireguards. On bushfires, Australia has as much to give the world as to get.

That extends to fire's cultural significance. We are a uniquely fire creature on a uniquely fire planet. Other animals knock over trees, dig holes in the ground, hunt, eat plants: we do fire. It is a distinctive signature of our ecological agency. Australia's intellectuals have long lamented their cultural isolation, not merely their geographic distance but the seeming remoteness of the country from the great questions about what it means to be human, and the monumental art and literature that such queries inspire. Yet its burning bush is an arena where Australians engage just such questions, for environmental ethics might well begin when early hominins picked up the firestick, the first of our Faustian bargains. Our decisions about what fires to apply and withhold continue to rely upon, and express, our core moralities, our sense of who we are and how we should behave. Australia can boast achievements in fire management equivalent to the Sydney Opera House in architecture. There is no reason not to have more. Those who would scoff at such enterprises as not truly cultural have a feeble, moth-eaten understanding of humanity. The manipulation of fire is as fundamental as it gets.

Three: Paradoxically, fire has not been at the core of fire management. Fire always seems at the fringe of other subjects, never the centre. This circumstance has a long pedigree. Earth, water, air —

all the ancient elements can boast their own academic disciplines; the only fire department on a university, however, is the one that sends emergency vehicles when an alarm sounds. This inability to create a pyro-centric paradigm has institutional consequences. The two prevailing paradigms for governing fire management, risk management and biodiversity (or ecological sustainability), are not grounded in fire but in the protection of their core assets from fire. Fire belongs on their periphery: they strive to keep it there.

Does this matter? Yes and no. Existing strategies can bolster the prevention of bad fires; they seem less eager to advance the cause of good ones. Both accentuate the prevailing public perception that, while fire may bring some benefits, it is predominantly a threat and a problem, and they reinforce technocratic instincts to resolve such dangers not with minute attention to the details of landscape but with complex managerial systems.

A doctrine of risk management lumps fire with all other hazards, under a common quadripartite formula that treats flame as though it were a cyclone or a toxic spill, and organises emergency response under an infinitely portable incident-management system. In the Alpine fires, country people often objected to what they saw as ignorant outsiders who blundered around the landscape unaware of the peculiarities of the scene, substituting mechanical muscle for hard-won local knowledge. But that is nothing compared to the practice of shunting officers trained in the incident-management system from Australasia to firelines in North America, as happened in 2000 and 2002, where they have even less understanding of local fuels, weather, indigenous customs, and jargon. (Americans have since come to Australia as well.) Whatever the exchanges contribute to trans-Pacific mateship, the enterprise testifies to the triumph of a managerial ethos over environmental particularities.[30]

Absorbing bushfire within a schema of risk management is not an Australian anomaly. It is not a case that the minds behind it are isolated, or quirky, or have suffered a kind of institutional sunstroke. They are riding a global trend that seeks to contain disasters within universal, bureaucratic-friendly formulas. All industrial nations, and cognate fire countries like Canada, are trending the same way; if anything, Australia is leading. Its long litany of catastrophic fires, its hard-won experience mustering bushfire brigades, and its tradition of careful scrutiny about how flames actually strike down houses and victims have made the transition to an all-hazards management model relatively simple and inevitable. The concern is that the prevailing paradigm has retained bushfire's standing as a disaster-in-the-making. Risk management has proved it can provide smarter social services; it had not yet demonstrated it can yield better land management.

Similarly, an emphasis on biodiversity shifts attention to the protection of ecological assets. Its advocates operate under a vision that, with cause, views Australia's environmental havoc as the result of people doing bad things, and does not appreciate how much value might have resulted from people doing good things. It is thus more concerned with stopping what it sees as pernicious burning than with encouraging beneficent burning. Besides, they reason, a message about the mixed virtues of anthropogenic fire might confuse the public. Moral clarity and the precautionary principle argue to shut off the smoke, if only through damning by faint praise. This, however, is precisely the attitude that, in inverse, underwrote imperial forestry's fire-prevention campaigns.

What matters here is that neither project grants special standing to fire. It is simply a threat for which universal formulas for management, properly executed, can lessen its impact. Of course it may have been a part of the Australian scene for eons, but that is no less true for other natural hazards that threaten lives, property, and biodiversity; the challenge in all these cases is to lessen their shock. Yet fire is arguably different; different in how it operates in nature, different in its relationship to humanity. So while a doctrine of 'risk mitigation', which seeks to replace 'hazard reduction', might save more houses from bad fires, it will not help get good fires back into the landscape. Besides, one might ask, à la the Esplin Inquiry, just what 'risk' is being mitigated?

Advocates of the new order will scoff; there can be no going back to the bad old ways; Australia has dallied too long in joining the modern world, even if that contemporary world is doing no better and perhaps doing more poorly with huge force and cost. This failure comes precisely because those institutions do not, or cannot, function at a landscape scale sufficiently fine-grained with the opportunism and nose-to-the-ground sensitivity to match the variety of the world under their care. The strategies are not informed by fire: fire is made to conform to other formulas. The proper core of fire management, however, should be fire, fire in the quotidian of routine life on the land. Mostly, we scrutinise fire on the fringe, but from perspectives on the outside looking in toward the fire; we need also to look outward from a vantage point within and centred on fire. The practical and conceptual core should be fire on the land.

Still, advocates for the contemporary approach may shake their heads and declare that there is no reason, in principle, why a commitment to risk management and biodiversity should compromise wise fire management. A betting man could find good odds, though, that in practice they will.

Four: Australians would do well to honour those who created, refined, and disseminated the Australian strategy because they kept a hand on the firestick and the firestick in the countryside. Rural burning acquired intellectual and institutional discipline, and fire remained. It was not always the right fire, and the practice certainly did not vanquish wild fire, but it has made possible the redirection of controlled burning to the new purposes Australians prefer for their public estate.

The identity politics of burning can be acrimonious and encumbered by collateral agendas; yet that does not change the fact that the achievement was genuine, and that the legatees of the Australian strategy are a reservoir of practical knowledge that Australia desperately needs to tap. Ignoring that knowledge is akin to extinguishing the firestick, for the firestick cannot maintain itself and has no genetic instinct or learned lore for how it should behave. The firestick is only as good as the person holding it. The Australia that is replacing hoofed beef and wool with biodiversity needs not only to grab that firestick but to be educated in its subtle deployment. Australia lost most of its Aboriginal fire lore; it cannot afford to lose the hard-won lore of forestry's fire-stick as well.

The four options for managing bushfire align with national or cultural models, what one might term the wilderness, aboriginal, cultivated, and fire-suppression models. Australia has need of all, adapted to particular places, but for its expansive reserved parks and preserves, the aboriginal model — with a small 'a' — would seem the most appropriate. What an aboriginal model is not is simple historical restoration, a reinstatement of the hypothetical fire regimes of pre-European contact. Nor does it seek to expand the dual-management regimes that exist at Kakadu National Park, where Aboriginal residents and park managers have established the terms for a partial reinstatement of traditional methods for a partial reclamation of traditional landscapes. That model will thrive in places, but cannot serve as the basis for a more national project.

Rather, an aboriginal strategy accepts that there is no return to the former ways. A true 'restoration' is impossible, and the mere attempt condemns the exercise to irony, of which fire history already has an over-abundance (I personally look forward to a post-ironic culture). An aboriginal model is not a recreation of the past but a reconstitution organised along analogous principles. Specifically, it points to fire management based on control over ignition.

It refers to a strategy in which the primary practices involve applying and withholding fire; not turning the process over to nature, not cultivating the fuels, not attempting to abolish burning, but by suitable use of the firestick seeking to protect against the fires you don't want and to promote the ones you do. The strategy differs from hazard-reduction burning in both its philosophy and practice. The object is not, as in the former, simply better fire protection but a proper habitat. The techniques are not variants of silviculture or suppression turned upside-down. Too much of prescribed fire is a set-piece, with its conditions established bureaucratically: a designated time and place, equipment for control on hand, and so forth. If any item on the checklist is not satisfied, the process can shut down.

But this is not, historically, how firestick farming operated. A modern analogue would resemble a kind of fire foraging, with people on the land matching available fuel with spark. It would involve a long season of burning, beginning with meagre, trivial burns and building as conditions mature. It would rely on varied means of burning, from spot ignitions to lines of fire, as circumstances and purposes warrant. It would be opportunistic, adapted not only to the evolving environmental details of a season but variances between years, with some years too wet to burn and some too dry. When conditions are right, you must move, which means you must be on the site and must recognise the occasion. Firestick ecology is labour intensive, although modern technology can substitute. Helicopters can disperse incendiaries on short notice and to specific locales; monitoring networks can congeal networks of information. What matters is the density of site-specific knowledge and the nimbleness to respond. While that does not demand firestick-wielding mobs roving through the country, it does require equivalents and surrogates, such as mechanical transport, densely coded digital knowledge, and ecological songlines.

Probably the most interesting exemplar today is Parks Canada, which is rapidly developing techniques for an aboriginal fire regime suitable to the Rocky Mountains and boreal forest. What drives the process administratively is a commitment to ecological integrity, not wilderness or heritage, and an administrative stipulation that each park establish a historic norm of burning and then burn at least half that amount. This fire quota is not a question of discretion or environmental noblesse oblige; it is a mandate, although one which only a handful of parks can at present achieve. What makes the experiment even more impressive is that most of these forests cannot be surface-burned, in the classic Australian way, for fire either smoulders in organic soils or erupts into and propagates through the canopies. Accordingly, Parks Canada sets prescribed crown fires. By close attention to fire behaviour, as well as using aircraft, they do so with minimal pre-treatments and small staffs.

The presumption is that they will burn unless unusual circumstances stop them, rather than that they will not burn unless special circumstances force them to do so. It's a model Australia might do well to ponder.

Finally: Amid our self-directed discoursing, we should remember that bushfire isn't listening. It doesn't hear our lecturing, blaming, scheming, haranguing. It can't parse our doctrinal prescriptions and clever managerial schemes. Our words are only so much vapour ready to be sucked into the next convective column. For all its symbiotic heritage with humanity, fire can exist by itself. Remove people, and fire will continue — not everywhere, and certainly not with the regimes that have characterised its geography these past tens of millennia, but flame will outlive us. We did not invent fire: we captured it from nature, and left untended, to nature it will return.

It has been said that the bushfire — the wild fire — will remain a part of Australia that may be forever alien; ineffable, unassimilable, an inextinguishable emblem of a nature beyond our will. Contemporary Australia will never absorb it nor wholly control it, and will probably not wish even to try. But Australians can negotiate a shared landscape; they can establish the terms of engagement; they can inscribe a text with the environmental grammar that bushfire understands, a biotic parchment of ignition and landscape, as read through wind and terrain. We can keep bushfire at bay by shaping the bush.

If land is the medium, the firestick is the means. In the great chain of combustion, the firestick has been the link between our wishes and nature's. It has morphed dramatically, from grasstree to Vesta match to diesel driptorch; the hands that grasped it have belonged to hunters, foragers, farmers, graziers, arsonists, campers, foresters, ecologists. So, too, have the purposes metamorphosed, from fire drive to green pick to hazard reduction. But throughout, the firestick has remained Australians' point of contact with their combustible bush. They put down that implement at their peril. Give up the firestick and you may not be able to speak to, much less to tame, bushfire at all. The only fire will be feral fire.

Where Australia sees the universe

All cultures have places where the sacred and the profane seem to meet, or where a portal opens between worlds. Traditionally, they are sites for temples, shrines, or rituals, or have been touched by myth or Dreaming. For societies shaped by the Enlightenment, they are often observatories, where technological wizardry assists scientific inquiry into the dynamics and origin of the cosmos and our place within it. For settler Australia this points to Mount Stromlo, the site of the country's oldest and most elaborate facility. Here, Australians can ask the universe — and, it would seem, the universe is prepared to reply.

Upon entering, a wall of posters and video displays informs visitors about the facility's exploration of the furthest frontiers and the deepest mysteries of the cosmos. Of particular pride are exhibits that describe recent work on interstellar gaseous blobs and clouds, and streams of gas jetting out of black holes. It is a facility committed to looking out, mapping the Southern sky, and yet in doing so, also looking in, plotting humanity's longing to know its place in the Great Scheme of Things. The instruments that scan the very borders of light itself also peer into the depths of the human soul. At such sites macrocosm and microcosm fuse.

On 18 January 2003 the cosmos returned that gaze. A conflagration powered by the volatilisation of the vast Australian bush blew over Mount Stromlo. Streams of combustible gases jetted out of pine plantations that clothed the Mount, building to a crescendo at its summit, amid billowing clouds of raw gases that could pass for nuclear pillars of fire in a far nebula. A successful evacuation ensured that no lives were lost, but the site was seared; only three out of a dozen buildings survived, and four telescopes melted, leaving husks that resembled sacked mosques; the immolated forest became a biotic black hole on the landscape, subsequently salvage-logged or scalped, piled, and burned. That spiral of flame turned the observatory's gaze inward, and in doing so it created a lens by which Australians, and humanity at large, might contemplate the elemental forces that sculpt the material world and might ponder their place within it. At that moment of impact, microcosm and macrocosm fused.

The resolve to rebuild was instinctive. Almost before the ashes cooled, even as a fight over insurance claims promised to be long and bloody, the governing institution, the Australian National University, announced it would replace the losses. Too much was at stake to abandon the facility: it was too deeply a part of international science and the heritage of Canberra, its loss too damaging to Australian prestige. Australia might fret over its global standing in the arts and humanities, but at Mount Stromlo it contributed boldly to the scientific quest to understand the ultimate standing of humanity in the cosmos. The telescopes would rise again.

The question about reconstructing the surrounding landscape

was trickier. The panorama from the summit encompassed three broad scenes. There were the belt of national parks and public lands along the mountains, the pastures and fields of a rural economy, and the dense forests of exotic *Pinus radiata* that dappled the countryside in thick patches. The parks, the paddocks, the plantations — the fire surge had struck them all, but with vastly different outcomes. The native bush rebounded briskly, stalled only by a stubborn drought. The pastoral country brushed off the burn, its primary losses in livestock, a resilience made possible in part because drought and overgrazing had shorn it of the woolly combustibles that fed the most ferocious flames. The pines, however, were immolated, and the landscape they once clothed, mauled. The scorched trunks were logged, or felled and dragged into windrows, the soils gouged and then shoved around to retard erosion. The densest woods had sparked the most savage burns.

There was reason to pause before reinstalling what might well lead at some future time to another fiery self-consummation. That had, in truth, happened once before when the pineries fell to the last regional conflagration, in February 1952. Then some outbuildings had burned, along with the workshop stocked with delicate and nearly irreplaceable machinery, and a desperate staff stood before the observatory itself with wet blankets and a sense of doom as the electricity and pumps failed. A light rain providentially dampened the flames and spared the facility. The workshop was rebuilt, the road reopened, and the blasted pine forest hauled away and replanted. At Mount Stromlo conflagrations were not a visitation from an alien galaxy but an emanation from within the landscape itself.

The choice of what to do next on those scalped slopes, of course, is for Australians to decide. But what grabbed my mind's eye was not the once-blackened trees or the now-stripped hillsides, but the ephemeral yet ineffable fire. That is surely a topic as worthy of hard scholarship as the deep forces hundreds of millions of light-years distant. The observatory's visitors' centre could do well to match its exhibit about the nuclear dynamics of nebulae with one about the combustion dynamics of the place from which we observe those scenes.

Flame is as elemental to Earth as gaseous clouds and dark matter are to the dynamics of the universe, and it is no less fundamental to humanity's place in the biosphere. The intellectual challenge of fire far exceeds that of stellar origins and extinctions. It is not simply that fire is a product of the living world, its chemistry and complexity far greater than anything astrophysics might conceive, but that fire, through humanity's monopoly over it, has entered a moral universe before which the most profound mathematics and complex instruments stand dumb. The character of future fires will depend on the character of the landscape, and that will depend on the character of the people who must choose what the place should be and what they can or can't do (or do by accident) regarding that decision. The cosmos of values is far more formidable than that cosmos of red giants, white dwarfs, and blue streamers about which astronomy can speak. There is not much we can do about the Creation. But with regard to landscapes and fire, we are a creator.

There are good reasons to point lenses to the heavens. There are also solid reasons to point them to the Earth. And where the macro and the micro merge into a single cosmos, there we might well want to pause and contemplate. Especially on a continent so prone to burning as Australia, where fire's removal can be as ecologically powerful as its presence, where no aspect of life is unaffected, where fires can roam on a scale visible from space, there is justification for an infrastructure of scholarship about free-burning fire equivalent to that at Mount Stromlo for the rhythms of stars. That process may have begun with the $100 million endowment for a Bushfire Cooperative Research Centre, belated recognition that the firestick is as powerful an instrument as the largest lens.

For bushfire, too, can be a place and a moment where the quotidian and the extraordinary meet. Around a billy fire people gather, against a wildfire they rally, and amid pyrocumulus plumes that tower into the sky they watch as flames become points where heaven and earth momentarily join in an immense burnt offering. With bushfires, Australians have a subject to which they have contributed with genuine force in the world. Through them Australians can see the universe — and, it would seem, the universe is prepared to reply. At Mount Stromlo in 2003 it replied in tongues of flame. In 2009 it spoke in Victoria, and in 2019, with inextinguishable flames and smoke that spread around the world, it screamed.

EPILOGUE

Black and forever

Neither the fires nor their political furore ended when the ashes of 2003 finally cooled. The Black Saturday bushfires of 2009 tested fire protection in Victoria beyond its elastic limits. The as-yet unnamed bushfires of 2019–2020 — spreading like a pyric plague — have changed the dynamics of confronting long-duration bushfire seasons, as they mutate from the shock-and-awe of an explosive blow-up to something like an endless siege that has driven the capabilities of fire agencies, like the residents of Balmoral and Mallacoota, to the beaches.

On 7 February 2009 bushfires blasted over Victoria — a classic fire-flume eruption. Of some 400 fires, 15 became major, and they killed 173 people, incinerated all or parts of communities, and stunned even a country that had throughout its settlement history has known and stoically named big burns. This time critics got the royal commission they craved, even before the fires were extinguished.

Two topics proved particularly contentious — one dealing with response, and the other with mitigation. The first centred on the stay-or-leave policy, which became a major post-burn casualty, in part because too often the preliminary part of that formula — prepare — got overlooked, and in part because the fires burned at intensities off the historic scales. The upshot was to instigate a catastrophic category of fire danger, a condition that warned a fire would likely range beyond the ability to fight or defend against. The second topic, unsurprisingly, flared over hazard reduction burning. A committee proposed to the Teague Commission that 10 per cent of public lands should become part of a regular burning regimen; this was watered down to a rolling average of 5 per cent. That amount has never been met — a lapse that to some observers was appropriate, and to others, a stigma of ignorance and incompetence. The Teague Commission was unable to resolve the issue — it was only able to restate it amid a renewed sense of urgency.

The narrative template also cracked. The magisterial omniscience of Judge Stretton broke into hundreds of voices, each with a story that needed to be heard. The commission struggled to muster them into a common institutional corral. But the tension between the individual and the collective intruded in revealing ways in the investigation over cause. Arson became a touchstone. Living in Australia had challenges, it was argued, but it was bad actors who turned tough circumstances into tragedy. Yet the problem was systemic. Powerlines kindled fires that killed 162 people — powerlines that emanated from the coal-fired Hazelwood Power Station (Australia's largest). In 2014 an ember storm ignited the Hazelwood mine itself. By such means flames arced between living and lithic landscapes.

Then an epidemic of bushfires began in September 2019. They visited every Australian state, though New South Wales and

Victoria were the most severely hammered, and were expected to burn through February 2020. The Australia Institute estimated that flame or smoke affected 57 per cent of Australians. The Forever fires kept Australia in world news, week after week. The politics promised to reach an early boiling point. There were mutterings about a federal royal commission, though there have been 57 official inquiries since Black Friday, and sceptics wondered what another could say that hadn't been said many times before. And, predictably, there were political lines drawn over what the role of deliberate burning should be.

What made the Forever fire season distinctive, however, was less its breadth across Australia than its siting within a global panorama. Alarms were sounding over fires in the boreal forest, in the Amazon and the Congo, and of course in California, Victoria's American cognate in fire, which was coming off a wave of serial conflagrations. California has a long chronicle of big burns, but four major outbreaks in three years was unprecedented. Several somethings were at work to alter the cadence and character of wildfire. Among them was climate change: global warming was acting as a performance enhancer on pre-existing dispositions to push the tempo, the vigour, and the meanness of landscape fires beyond historic boundaries.

Climate wasn't the only culprit, but fossil fuels largely were because they also affected how people lived on the land and how they applied and withheld fire in the bush. Like first-contact firesticks working best on already fire-disposed places, so the Pyrocene will express itself initially and most dramatically on those places that have a deep history of burning. Australia can't contain climate change on its own, but it can affect how an unhinged climate expresses itself through flame. The curious career of the firestick is far from finished.

The Still-Burning Bush ends formally with the Alpine fires. Its central themes, however, persist through the 2009 and 2019–2020 seasons. But there is one critical distinction between the era of Black Friday and that of Black Saturday. The earlier fires were Australian affairs, produced by and played out in Australia. The new fires have a global reach. They have attracted worldwide interest, and they appear to integrate planetary causes, especially the phase change in earthly fire history prompted by the pyric transition. Previously, Australia had endured fire crises. Now it appears to be on the leading edge of a fire epoch.

The endless burning seemed to reach Biblical proportions. Those who distrusted the science might turn to scripture instead. Proverbs 30:15-16 declared that there are four things that are never satisfied. 'The grave; and the barren womb; the earth that is not filled with water; and the fire that saith not, It is enough.' Australia had three of them.

Acknowledgements

This reissue is the brainchild of Henry Rosenbloom, who had my thanks for publishing the first edition, and has it now, doubly, for the second.

The original project was the outcome of a month's fellowship at the History Department, Research School of Social Sciences, Australian National University, in 2005. The invitation came from Tom Griffiths, and I'm pleased to again thank him, his wife, Libby Robin, and the always-interesting students in the program for a delightful visit. As I informed my hosts, it was not really necessary to burn a bit of Black Mountain as a greeting, but I appreciated the gesture. Special thanks to Susan Mary and Robert Withycombe for a day's trek to Mount Stromlo and other scenes from the 2003 fires. I thank, too, those fire colleagues I had a chance to chat with while in Canberra: Malcolm Gill, Phil Cheney, Athol Hodgson, and Peter Kanowski. From all I learned much, and found much to admire, but to state the obvious, the sum of my interpretation is not theirs.

Along with the new preface and the new epilogue, I have new acknowledgements as well. I wish to extend special thanks to Edwina Johnson and Rosemary Sorensen, respectively the organisers of the Byron Bay and Bendigo writers festivals, for an invitation to participate in August 2019, and to Katie Holmes for organising talks at La Trobe University between those events. The occasion for the invitations was the tenth anniversary of Black Saturday. Over the years I had wandered far from Australia. Fortuitously, the festivals and lectures repositioned me to follow the 2019–20 bushfire season. I had barely shaken off jetlag when the first sparks kindled.

Notes

1 See, for example, Richard Wrangham, *Catching Fire: how cooking made us human* (New York: Basic Books, 2009)
2 D.M.J.S. Bowman, *Australian Rainforests: Islands of Green in a Land of Fire* (Cambridge University Press, 2000). See also his careful scrutiny, 'Tansley Review No. 101. The impact of Aboriginal landscape burning on the Australian biota', *New Phytologist* 140 (1998), pp. 385–410
3 Gifford H. Miller, et al, 'Ecosystem Collapse in Pleistocene Australia and a Human Role in Megafaunal Extinction', *Science* 309 (8 July 2005), pp. 287–90. Quote: Christopher N. Johnson, ' The Remaking of Australia's Ecology', *Science* 309 (8 July 2005), p. 256
4 Sylvia J. Hallam, 'Fire and Hearth. A Study of Aboriginal Usage and European Usurpation in South-western Australia', *Australian Aboriginal Studies* No. 58 (Canberra: Australian Institute of Aboriginal Studies, 1979), preamble
5 Rhys Jones, 'Fire-Stick Farming', *Australian Natural History* (Sept 1969), pp. 224–28. Bill Gammage offers a magisterial expansion on the notion with his *The Biggest Estate on Earth* (Sydney: Allen & Unwin, 2012
6 Stephen J. Pyne, 'Fire Flume (Australia),' *World Fire* (New York: Holt, 1995), pp. 34–35

7 D.E. Hutchins, *A Discussion of Australian Forestry* (Perth, 1916)
8 Stephen J. Pyne, *Burning Bush: a fire history of Australia* (New York: Holt, 1991), p. 258; Brandis quoted in Stephen J. Pyne, *Vestal Fire: an environmental history, told through fire, of Europe and Europe's encounter with the world* (Seattle: University of Washington Press, 1997), p. 489
9 Pyne, *Burning Bush*, p. 258
10 Tiger episode quoted from Pyne, 'Nataraja', *World Fire*, p. 164
11 Rudyard Kipling, 'In the Rukh', *The Jungle Books* (Oxford: Oxford University Press, 1992), Appendix A
12 E.O. Shebbeare, 'Fire Protection and Fire Control in India', Third British Empire Forestry Conference (Canberra, 1928), p. 1
13 John Wesley Powell, *Report on the Lands of the Arid Region of the United States*, 2nd ed. (Cambridge: Harvard University Press, 1962, reprint), pp. 24–29, 113, and *Seeing Things Whole: the essential Wesley Powell*, ed. William deBuys (Washington: Island Press, 2001), p. 291, which reprints the 1890 article. For an alternative interpretation of Powell's role, see Andrew Denny Rodgers III, *Bernhard Fernow: a story of North American forestry* (New York: Hafner, 1968), p. 154
14 From Plumas Boundary Report of 1904, quoted in Louis Barrett, 'A Record of Forest and Field Fires in California from the Days of the Early Explorers to the Creation of the Forest Reserves' (U.S. Forest Service, 1935), p. 48
15 For the politics behind the controversy, see Stephen J. Pyne, *Year of the Fires: the story of the great fires of 1910* (New York: Viking, 2001)

16 The fullest summary of developments can be found in Stephen J. Pyne, *Fire in America: a cultural history of wildland and rural fire* (Seattle: University of Washington Press, 1995; reprint of 1982 edition), pp. 100–122. See also Ashley L. Schiff, *Fire and Water: scientific heresy in the Forest Service* (Cambridge: Harvard University Press, 1962) for a fascinating study of the internal quarrel between state-sponsored forestry and the science it supported.

17 S.B. Show and E.K. Kotok, "The Role of Fire in the California Pine Forests," U.S. Department of Agriculture, Department Bulletin No. 1294 (Government Printing Office, 1924), p. 47

18 Aldo Leopold, "'Piute Forestry' vs. Forest Fire Protection," in David E. Brown and Neil B. Carmony, eds., Aldo Leopold's Southwest (Albuquerque: University of New Mexico Press, 1990), p. 141

19 Pyne, *Burning Bush*, p. 26

20 Jones quoted in Pyne, *Burning Bush*, p. 256

21 Kessell quotes from Pyne, *Burning Bush*, pp. 267–68

22 Kessell quotes, from Pyne, *Burning Bush*, p. 268

23 H.G. Wells, *Travels of a Republican Radical in Search of Hot Water* (Harmondsworth: Penguin, 1939), p. 46. For a thoughtful synopsis on Stretton, see Tom Griffiths, 'Judge Stretton's Fires of Conscience', *Gippsland Heritage Journal* No. 26 (2002), pp. 9–18

24 National Interagency Fire Center, website: http://www.nifc.gov/stats/prescribedfirestats.html [accessed 6 January 2006]

25 Australian Supreme Court, 'The Queen v. Coroner Maria Doogan', ACTSC 74 (5 August 2005), para 190

26 Bruce Esplin, Malcolm Gill, Neal Enright, *Report of the Inquiry into the 2002–2003 Victorian Bushfires* (Abbotsford: State Government of Victoria, 2003), p. iii. Hereafter referred to as Esplin Report.

27 Forest Fire Victoria, *The Facts Behind the Fires* (May 2005). Many months previous to this study, the organisation had asked me join, which I did because I knew and liked some of the members, respected their collective fire wisdom, and was pleased to see fire policy discussed more openly. I was, however, uninvolved in any way with the report. Since then I have withdrawn, not from any disagreement over purposes, but from the belief that I can better voice my own concerns by being unaffiliated altogether.

28 *Esplin Report*, p. 75

29 *Esplin Report*, p. 74

30 For an example of local outrage, see The Eureka Project, 'A Case of Burning Neglect. A report from North-East Victorians regarding the 2003 bushfire crisis. 30 April 2003'. For the trans-Pacific exchanges, see Andrew J. Patrick, 'The Globalization of Wildfire. A History of the Australia/New Zealand Deployment at the Fires of 2000' (National Interagency Fire Center, 2001). For a good digest of the technocratic perspective, which includes indigenous knowledge but manages to shut out rural lore, see Geoffrey Cary et al (eds), *Australia Burning: fire ecology, policy and management issues* (Collingwood: CSIRO Publishing, 2003).

Sources and further reading

The good news is that the literature on Australian fire is large — among the largest in the world — and growing rapidly. That is also the bad news. What follows are full sources for those works I have cited or relied upon for this particular project, which is, in the end, an interpretive essay, not a work either of full synthesis or of archival scholarship. For background reading, I refer the reader to A. Malcolm Gill et al, eds., *Fire and the Australian Biota*; A. Malcolm Gill and I.R. Noble, *Bibliography of Fire Ecology in Australia*; and my own *Burning Bush: a fire history of Australia*, which carries the story up through 1983. Since I have lightly extended the range of the text, I have also updated sources, particularly regarding the Black Saturday fires.

Abbott, Ian and Neil Burrows, eds. *Fire in Ecosystems of South-West Western Australia: impacts and management* (Leiden: Backhuys Publishers, 2003)

Australian Supreme Court. 'The Queen v Coroner Maria Doogan; ex parte Peter Lucas-Smith & ORS, and The Queen v Coroner Maria Doogan and ORS; ex parte Australian Capital Territory [2005],' ACTSC 74 (5 August 2005)

Blainey, Geoffrey. *The Triumph of the Nomads*, rev. ed (Sun Books, 1983).

Bowman, D.M.J.S. *Australian Rainforests. islands of green in a land of fire* (Cambridge: Cambridge University Press, 2000)

_____. 'Tansley Review No. 101. The impact of Aboriginal landscape burning on the Australian biota,' *New Phytologist 140* (1998), pp. 385–410

Bradstock, R.A. et al, eds. *Flammable Australia. the fire regimes and biodiversity of a continent* (Cambridge: Cambridge University Press, 2002)

_____. *Flammable Australia. fire regimes, biodiversity and ecosystems in a changing world* (Collingwood: CSIRO, 2012)

Cary, Geoffrey et al, eds. *Australia Burning: fire ecology, policy and management issues* (Collingwood: CSIRO, 2003)

DeBuys, William, ed. *Seeing Things Whole: the essential Wesley Powell* (Washington: Island Press, 2001)

Dyer, Rodd et al, eds. *Savanna Burning: understanding and using fire in Northern Australia* (Darwin: Tropical Savannas Cooperative Research Centre, 2001)

Ellis, S. et al. *National Inquiry on Bushfire Mitigation and Management* (Canberra: Council of Australian Governments, 2004)

Esplin, B., A.M. Gill, and N. Enright. *Report of the Inquiry into the 2002–2003 Victorian Bushfires* (Melbourne: State Government of Victoria, 2003)

Eureka Project. 'A Case of Burning Neglect: a report from North East Victorians regarding the 2003 bushfire crisis' (The Eureka Project, 2003)

Fire Ecology Working Group. *Analysis of Disturbance by Fire on Public Land in Victoria* (Department of Natural Resources and Environment and Parks Victoria, 2002)

Flannery, Timothy. *The Future Eaters: an ecological history of the Australasian lands and people* (London: Martin Secker and Warburg Ltd, 1996)

Ford, Julian, ed. *Fire Ecology and Management in Western Australian Ecosystems.* WAIT Environmental Studies Group Report No. 14 (Western Australia Institute of Technology, 1985)

Franklin, Roger, *Inferno: the day Victoria burned* (Docklands: Slattery Media Group, 2009)

Fraser, Peg, *Black Saturday: not the end of the story* (Clayton: Monash University Publishing, 2018)

Gammage, Bill. 'Australia Under Aboriginal Management,' 15th Barry Andrews Memorial Lecture (Canberra: University College, 2002)

_____. *The Biggest Estate on Earth: how Aborigines made Australia* (Sydney: Allen & Unwin, 2012)

Gill, A. Malcolm, et al, eds., *Fire and the Australian Biota* (Canberra: Australian Academy of Sciences, 1981)

_____, et al. *Beyond the Bushfires 2003: environmental issues in the Australian Alps* (Australian Alps National Parks, 2004)

_____ and Peter H.R. Moore, 'Fire Situation in Australia,' *International Forest Fire News No. 26* (January 2002), pp. 2–8

_____ and I.R. Noble, *Bibliography of Fire Ecology in Australia* (Bushfire Council of New South Wales, 1986)

Griffiths, Tom. *Forests of Ash: an environmental history* (Cambridge: Cambridge University Press, 2001)

_____. 'Judge Stretton's Fires of Conscience,' *Gippsland Heritage Journal No. 26* (2002), pp. 9–26

Hallam, Sylvia J. *Fire and Hearth: a study of Aboriginal usage and European usurpation in south-western Australia*, Australian Aboriginal Studies No. 58 (Canberra: Australian Institute of Aboriginal Studies, 1979)

Hansen, Christine and Tom Griffiths, *Living With Fire: people, nature and history in Steels Creek* (Collingwood: CSIRO, 2012)

Hogg, Doris. '1952 Mt Stromlo Fire Remembered,' ANU Research School of Astronomy and Astrophysics [http://www.mso.anu.edu.au/info/fire/1952_fire.ph]

Hoggett, Jim and Aled Hoggett. 'When Will We Ever Learn?' Institute for Public Affairs *Backgrounder 16*(2) (May 2004)

Hooper, Chloe. *The Arsonist: a mind on fire* (Sydney: Penguin RandomHouse Australia, 2018)

Horton, D.R. *The Pure State of Nature: sacred cows, destructive myths, and the environment* (St Leonards, NSW: Allen & Unwin, 2000)

Joint Select Committee on Bushfires. *Report on the Inquiry into the 2001–2002 Bushfires. Together with the Minutes of Proceedings* (Parliament, NSW, 2002)

Jones, Rhys. 'Fire-stick Farming,' Australian Natural History 16 (1969), pp. 224–28

Johnson, Christopher N. 'The Remaking of Australia's Ecology,' *Science 309* (8 July 2005), pp. 255–56

Jurkis, V. et al, 'Fire Management in Australia: the Lessons of 200 Years,' in: *Joint Australia and New Zealand Institute of Forestry Conference Proceedings, 27 April–1 May 2003, Queenstown, New Zealand* (Wellington: Ministry of Agriculture and Forestry, 2003), pp. 353–68

Kanowski, P. J. et al. 'Inquiries following the 2002–2003 Australian bushfires: common themes and future directions for Australian bushfire mitigation and management,' *Australian Forestry 68*(2) (2005), pp. 77–87

Kohen, James. *Aboriginal Environmental Impacts* (Sydney: University of New South Wales Press, 1995)

Lewis, Henry and Theresa Ferguson, 'Fire Corridors, Fire Yards, Fire Mosaics: how to burn a boreal forest', *Human Ecology* 16, pp. 57–77

McLeod, R. *Inquiry into the Operational Response to the January 2003 Bushfires in the ACT* (Canberra: ACT Government, 2003)

Miller, Gifford H. et al, 'Ecosystem Collapse in Pleistocene Australia and a Human Role in Megafaunal Extinction,' *Science 309* (8 July 2005), p. 287–90

Murray, Robert and Kate White, *State of Fire: a history of volunteer firefighting and the Country Fire Authority in Victoria* (North Melbourne: Hargreen Publishing Company, 1995)

National Academies Forum. 'Fire! The Australian Experience. Proceedings of the 1999 Seminar' (Carlton: Snap Printing, 2000)

Neale, Timothy et al, 'Walking Together: a decolonising experiment in bushfire management on Dja Dja Wurrung country,' *Cultural Geographies 26*(3) (2019): 341–59.

New South Wales Rural Fire Service. 'Fire! The Australian Experience,' 2nd ed (NSW Rural Fire Service, 1998)

Patrick, Andrew J., 'The Globalization of Wildfire: a history of the Australia/New Zealand deployment at the fires of 2000' (Boise, Idaho: National Interagency Fire Center, 2001)

Petris, Stephen. 'A Review of All State and Federal Reports on Major Conflagrations in Australia During the Period 1939–1994' (Australian Fire Authorities Council, 1995)

_____ and Paul Potter. ' National Bushfire Preparedness Strategy' (Australian Fire Authorities Council, 1995)

Powell, John Wesley. *Report on the Lands of the Arid Region of the United States*, 2nd ed (Cambridge: Harvard University Press, 1962, reprint)

Pyne, Stephen J. *Fire in America: a cultural history of wildland and rural fire* (Princeton: Princeton University Press, 1982)

_____. *Burning Bush: a fire history of Australia* (New York: Holt, 1991)

_____. *Vestal Fire. an environmental history, told through fire, of Europe and Europe's encounter with the world* (Seattle: University of Washington Press, 1997)

_____. *Fire: a brief history*, 2nd ed (Sydney: NewSouth Publishing, 2020)

Rodgers III, Andrew Denny. *Bernhard Fernow: a story of North American forestry* (New York: Hafner, 1968)

Rose, Deborah Bird, ed. *Country in Flames. Proceedings of the 1994 Symposium on Biodiversity and Fire in North Australia*, Biodiversity Series, Paper No. 3 (Canberra and Darwin: Australian National University and Department of the Environment, Sport and Territories, 1995)

Russell-Smith, Jeremy, Peter Whitehead, Peter Cooke, eds. *Culture, Ecology and Economy of Fire Management in North Australian Savannas: rekindling the Wurrk tradition* (Canberra: CSIRO Publishing, 2010)

Schiff, Ashley. *Fire and Water: scientific heresy in the forest service* (Cambridge: Harvard University Press, 1962)

Select Committee into the Recent Australian Bushfires, House of Representatives. *A Nation Charred: report on the inquiry into bushfires* (Canberra: Parliament of the Commonwealth of Australia, 2003)

Stanbury, Peter, ed. *Bushfires. their effect on Australian life and landscape* (Sydney: Macleay Museum, 1981)

Teague, Hon. Bernard, Ronald McLeod, Susan Pascoe, *2009 Victorian Bushfires Royal Commission. Final Report. Summary* (Government Printer for the State of Victoria, July 2010)

Ward, David. 'Trouble in the Tuart. A Brief Fire History' (Perth: Department of Conservation and Land Management, 2000)

Ward, David and Rick Sneeuwjagt, 'Believing the Balga,' *Landscope, Fire: The Force of Life.* Special Fire Edition (Perth: Department of Conservation and Land Management, 2000)

Wells, H.G., *Travels of a Republican Radical in Search of Hot Water* (Harmondsworth: Penguin, 1939)